Theory and Practice of "Lucid Waters and Lush Mountains Are Invaluable Assets" in China

Huiyuan Zhang · Yujie Liu · Chao Zhang · Haiguang Hao

Theory and Practice of "Lucid Waters and Lush Mountains Are Invaluable Assets" in China

CEPG

Springer

Huiyuan Zhang
Chinese Research Academy
of Environmental Sciences
Beijing, China

Yujie Liu
Chinese Research Academy
of Environmental Sciences
Beijing, China

Chao Zhang
Chinese Research Academy
of Environmental Sciences
Beijing, China

Haiguang Hao
Chinese Research Academy
of Environmental Sciences
Beijing, China

ISBN 978-981-97-7339-8 ISBN 978-981-97-7340-4 (eBook)
https://doi.org/10.1007/978-981-97-7340-4

Jointly published with China Environment Publishing Group Co., Ltd.

The print edition is not for sale in China (Mainland). Customers from China (Mainland) please order the print book from: China Environment Publishing Group Co., Ltd.
ISBN of the Co-Publisher's edition: 978-7-5111-4957-2

This Springer imprint is published by the registered company Springer Nature Singapore Pte Ltd.
The registered company address is: 152 Beach Road, #21-01/04 Gateway East, Singapore 189721, Singapore

Editorial Board of the Collection

Editorial Board

Preface

The concept of "Lucid Waters and Lush Mountains are Invaluable Assets" forms a symbolic viewpoint and conclusion of Xi Jinping Thought on Ecological Civilization, representing a significant theoretical innovation proposed by Chinese President Xi Jinping after deep reflection on the relationship between development and protection. It also demonstrates a major theoretical achievement in adapting Marxism to the Chinese context. Furthermore, the concept of "Lucid Waters and Lush Mountains are Invaluable Assets" profoundly reveals the dialectical relationship and internal law between humankind and nature, and between economic development and environmental protection. Finally, it also provides a strong theoretical support, a direction, and a guideline in constructing an ecological civilization, a beautiful China, and realizing the sustainable development of the Chinese nation. In light of all the above, the concept therefore holds great theoretical, practical, historical, and global significance.

Through constant enrichment and development, the concept of "Lucid Waters and Lush Mountains are Invaluable Assets" has evolved into a comprehensive set of logically rigorous and science-based theoretical systems. These systems have then been turned into consensus reached by and actions taken by the Communist Party of China (the Party, or CPC) and society as a whole. In April 2021, the general offices of the Communist Party of China Central Committee and the State Council issued the *Opinions on Establishing and Improving the Value Realization Mechanism of Ecological Products*. This document stipulated that "under the premise of strictly protecting the ecological environment, diverse methods and approaches are encouraged in order to scientifically and rationally help realize the value of ecological products." This provided a new requirement and direction which helped comprehensively fulfill the concept of "Lucid Waters and Lush Mountains are Invaluable Assets," exploring a path toward this concept, and establishing and improving the value realization mechanism of ecological products.

In order to better practice the concept of "Lucid Waters and Lush Mountains are Invaluable Assets," the Ministry of Ecology and Environment of the People's Republic of China (MEE) has been launching campaigns to establish innovation bases on this concept since 2017 under the guidance of Xi Jinping Thought on Ecological Civilization, which are designed to explore new systems and practical actions that allow for ecological environmental protection at the same time as social and economic development. So far, 136 bases have been set up, all of which offer inspiring examples of how to fulfill the concept of "Lucid Waters and Lush Mountains are Invaluable Assets," as well as setting the stage for other areas across the country to follow. Regions all over China have proactively pushed forward the practice of "Lucid Waters and Lush Mountains are Invaluable Assets," making impressive progress and achieving results in the supply and value realization of ecological products, transformation paths and modes of the practice, and the innovation of long-term guarantee systems and mechanisms.

Green development has become the defining theme of social development. For the purposes of summing up the progress and outcome of practicing the concept of "Lucid Waters and Lush Mountains are Invaluable Assets" across China, studying its effects, evaluating its construction, and publicizing typical cases and experiences of the concept, we compiled the book *Theory and Practice of "Lucid Waters and Lush Mountains Are Invaluable Assets" in China*. This book mainly analyzes the connotations and formation processes of the concept of "Lucid Waters and Lush Mountains are Invaluable Assets" and its significance in this era. It also summarizes the progress, effect, and transformation modes of the practice of "Lucid Waters and Lush Mountains are Invaluable Assets," puts forward approaches for evaluating its construction effects of "Lucid Waters and Lush Mountains are Invaluable Assets," and on this basis, introduces ten practice cases which exemplify the "Lucid Waters and Lush Mountains are Invaluable Assets" concept.

Going forward, we will continue to enhance the practical exploration and theoretical innovation of the "Lucid Waters and Lush Mountains are Invaluable Assets" concept, summarize and produce transformation paths and modes of "Lucid Waters and Lush Mountains are Invaluable Assets," and gradually conduct evaluation of "Lucid Waters and Lush Mountains are Invaluable Assets" indicators in light of a new stage of new development. On the basis of implementing a new development philosophy and serving the establishment of a new development paradigm, we will support endeavors to make new progress in ecological conservation and build a beautiful China.

Beijing, China

Huiyuan Zhang
Yujie Liu
Chao Zhang
Haiguang Hao

Contents

Chapter 1
Research Background

1.1 The Increasingly Tense Relationship Between Economic Growth and Environmental Protection

1.1.1 An Increasingly Severe Ecological Environment Landscape Around the World

With the advancement of global industrialization and urbanization, mankind's demand for natural resources has increased dramatically; resources such as minerals, forests and water bodies have been exploited on a large scale; natural ecosystems such as forests, grasslands and wetlands have been seriously damaged; and a large number of pollutants have been discharged into the environment, exceeding the carrying capacity of nature and leading to serious environmental pollution, which in turn threatens the health and survival and development of mankind. Eight major pollution incidents occurring between the 1930s and the 1960s all dealt a heavy blow to the ecological environment and public life.

Since this period, people have begun to reflect on the relationship between humankind and nature. However, environmental pollution, biodiversity loss, ecological damage and other issues arising from human activities remain severe—environmental pollution, global warming and biodiversity loss have become the top three environmental issues in the world. The conflict between economic growth and environmental protection has evolved from regional issues to global contradictions.

According to statistics from the United Nations Environment Programme (UNEP), from 1970 to 2020, the world's population doubled, economic aggregate grew nearly fivefold, chemical products, waste and pollution increased greatly, greenhouse gas (GHG) emissions grew onefold, and annual deaths due to air pollution approached 7 million. Studies from the UN Intergovernmental Panel on Climate Change (IPCC) show that as a result of GHG emissions, the world's surface temperature was 1.09 °C higher from 2011 to 2020 than that recorded from 1850 to 1900, close to the 1.5 °C

H. Zhang et al., *Theory and Practice of "Lucid Waters and Lush Mountains Are Invaluable Assets" in China*, https://doi.org/10.1007/978-981-97-7340-4_1

threshold. If this threshold is reached, the world is more likely to see phenomena such as the disappearance of Arctic sea ice, large-scale extinction of coral reefs and thawing of methane-rich permafrost, as well as permanent changes to the ecological system of the earth (IPCC 2021). The latest evaluation released by the World Wildlife Fund (WWF) and the Zoological Society of London (ZSL) in 2020 shows that the number of monitored wildlife populations has decreased by two thirds on average since 1970 (Almond et al. 2020). It is generally believed by the scientific community that this is the biggest biodiversity crisis since the extinction of dinosaurs 65 million years ago, which was caused by human pollution and destruction of the ecological system, which is the natural habitats of species (Li Gang 2021). Global warming will also seriously exacerbate the biodiversity crisis, and its impact will only grow.

Given the economic development needs and a rapidly growing population, humanity is still faced with problems resulting from our overuse of limited resources. Studies indicate that from 1992 to 2017, the quantity of per capita fresh water available in the world fell by 26%; "dead zones" in oceans where organisms cannot survive due to pollution and low oxygen levels increased by 75%; forestry areas were down by nearly 300 million acres for the purposes of making way for farmland; while with biodiversity decline, the number of mammals, reptiles, amphibians, fish and birds dropped by 29% (Ripple et al. 2017). A study entitled "Where Might We Find Ecologically Intact Communities?" shows that most terrestrial habitats on earth have lost ecological integrity, and less than 3% of the earth's land can be considered ecologically intact, with the same fauna and flora that were found 500 years ago, in pre-industrial times, before major human impact occurred (Plumptre et al. 2021).

1.1.2 Environmental Issues in China's Development Process

As its economy and society develop, China has also been confronted with severe ecological environmental issues, and hence started to embark on a road in pursuit of coordinated development of the economy and ecological environment.

In terms of ecological security, the pattern and service functions of China's ecological system have generally improved since 2000, but traditional water and soil erosion, stony desertification, and desertification have remained severe. Ecological quality is still poor, while degradation of watershed ecology and urban ecological functions, and worsening of human settlement caused by urbanization and resource development have increased, according to findings from the "Remote Sensing Investigation and Assessment on Ecological Environmental Changes in China from 2000 to 2010" project (Ouyang Zhiyun 2017). The Plan for Ecological Environmental Protection in the 13th Five-Year Plan Period (2016–2020) states that wetland areas in China have decreased by around 5.1 million mu (or about 340,000 ha) on an annual basis in recent years, meaning that the survival of over 900 species of vertebrates and over 3700 species of higher plants is under threat. Over-exploitation and utilization of resources has caused huge ecological damage, with constant encroachment on ecological spaces, and in some regions, severe destruction of ecological resources, adding to difficulties in ecosystem protection.

With regards to environmental pollution, emissions of chemical oxygen demand, sulfur dioxide and other major pollutants in China stood at around 20 million tons before the 13th Five-Year Plan period, with environmental carrying capacity (ECC) approaching or exceeding the upper limit. At that time, 78.4% of cities failed to meet the standards for air quality, and the proportion of heavily polluted days strongly reflected by members of the public accounted for 3.2%, while some regions were particularly prone to heavy air pollution in winter. Some areas suffered from poor access to safe drinking water, the pollution discharge pattern did not match with the carrying capacity of the water environment, many black and odorous water bodies were present in urban built-up areas, the issue of eutrophication of lakes and reservoirs remained serious, and some river basins suffered from severe pollution. China also faced the problem of soil pollution. According to a national soil survey published in 2014, 16.1% of China's soil and 19.4% of its arable land (proportion of soil survey points with contamination above maximum tolerance) showed contamination. Furthermore, a large gap in the quality of environmental public services between urban and rural areas made governance and improvement tasks daunting.

1.2 The Global Challenge of How to Coordinate Development and Protection

At their roots, environmental problems are problems with the ways in which we live and develop. The American economists Grossman and Krueger studied the relationship between environmental quality and economic development, pointing out that when a country's economic development level is low, environmental pollution is mild. But as per capita income increases, the environment worsens, and when economic development reaches a critical point, environment pollution starts to be improved alongside the increase in per capita income (Grossman 1993). However, the increasingly prominence of ecological environmental issues has proved that solving these issues by improving economic development comes at a greater price, and can even lead to irreversible consequences. The development mode of treatment after pollution can no longer adapt to the current reality of globalized ecological environmental issues (Pan Yue 2018).

Economic development and environmental protection are two sides of the same coin, and it is therefore necessary to analyze their relationship in a dialectical way. Irrational and inefficient resource utilization in the process of economic development naturally triggers ecological environmental issues, while a decline in ecological environment quality inevitably affects improvements in economic development. Soley considering environmental protection or economic development is therefore not the best solution. Coordinating economic development and environmental protection is crucial to tackling current ecological environmental issues, calling for the joint efforts of all countries (Xi Jinping 2019). The UNEP indicates that a stable climate and a healthy ecological system matter a great deal to the economic recovery and sustainable development of all countries, all of which should prioritize low-carbon energy, green transportation, green building upgrading and energy efficiency, as well as natural capital and the research and development of green technology, with a view to driving global green recovery (UNEP 2021).

The Chinese government has also attached great importance to green development and sustainable development. The 18th CPC National Congress integrated the building of ecological civilization into the "Five-sphere Integrated Plan" of socialism with Chinese characteristics. The CPC Central Committee and the State Council then released the *Guidelines on Accelerating Ecological Civilization* and *the Integrated Reform Plan for Promoting Ecological Civilization*, while the Fifth Plenary Session of the 18th CPC Central Committee adopted green development as one of the "Five Major Principles of Development" guiding China's economic and social development. President Xi Jinping has emphasized on many occasions that lucid waters and lush mountains are invaluable assets, and that to protect the environment is to protect productivity. Advancing green development and building ecological civilization has become a fundamental philosophy for China's economic and social development in the 13th Five-Year Plan period as well as for years to come, and also an important guiding principle and mission in socialist modernization with Chinese characteristics (Zhang Qiang et al. 2015).

The 19th CPC National Congress made a comprehensive summary of and important arrangements for the building of ecological civilization. It elevated building ecological civilization as a "millennium program", integrated "beautiful" into China's modernization goals, and included "high-quality ecological products" into the scope of people's well-being. The Fifth Plenary Session of the 19th CPC Central Committee particularly discussed promoting green development and the harmony between man and nature, noting that China should follow the concept of "Lucid Waters and Lush Mountains are Invaluable Assets", work to respect nature, follow its ways, and protect it, prioritize resource conservation and environmental protection and let nature restore itself, and maintain the boundaries of natural security. China should also deeply implement sustainable development strategies, improve planning and coordination mechanisms in the field of ecological civilization, shape the ecological civilization regime, promote the all-round, green transition of economic and social development, and pursue modernization featuring harmony between man and nature.

1.3 Relevant Studies on Coordinating the Relationship Between Economic Development and Environmental Protection

1.3.1 Studies on Sustainable Development

Sustainable development represents a development mode universally accepted by the international community, coordinating the relationship between the three systems of economy, society and environment (CRCSD 1994). In 1972, the United Nations convened the United Nations Conference on the Human Environment in Stockholm, the capital of Sweden, and released the *Declaration of the United Nations Conference on the Human Environment*, calling on governments and peoples of all countries to jointly maintain and improve the human environment, for the benefit of people around the world and future generations. This was the first time in human history that intergovernmental organizations jointly discussed issues involving environment and development, providing a building block for the proposal of sustainable development.

In 1980, the *World Conservation Strategy* drafted by the International Union for Conservation of Nature (IUCN) under the entrustment of the UNEP adopted the concept of "sustainable development" for the first time.

In 1987, the World Commission on Environment and Development (WCED) submitted the report *Our Common Future* to the United Nations. The report stated that it is necessary to find a development road that meets the needs of the present without compromising the ability of future generations to meet their own needs, and balances resource and environmental protection and economic and social development. The report proposed a definition of sustainable development for the first time, while clarifying the principles, requirements, objectives and strategies of sustainable development.

In 1992, the United Nations Conference on Environment and Development (UNCED) was held in Rio de Janeiro, Brazil, where the *Agenda 21* was released. The *Agenda* stated that it is necessary to make a significant change to the traditional development pattern of "high production, high consumption and high pollution" and the concept of "treatment after pollution" since the Industrial Revolution, and to embark on a whole-new sustainable development road. This marked the shift in sustainable development from theory to practice.

In 2002, the World Summit on Sustainable Development (WSSD) was held in Johannesburg, South Africa, where representatives from 192 countries, including heads of state and government of 104 countries, gathered together to explore future plans for global sustainable development. Representatives endorsed the Johannesburg Declaration on Sustainable Development and the Plan of Implementation, which made clear the blueprint of global sustainable development in the coming 10 to 20 years covering environmental protection, poverty elimination and prosperity advancement, and proposed new development goals and timeline concerning the social, economic and environmental issues of sustainable development.

In 2012, world leaders met together in Rio de Janeiro, Brazil for the United Nations Conference on Sustainable Development (UNCSD), primarily aiming to reach a new political commitment on sustainable development, evaluate progress in delivering existing commitments and implementation gaps, and jointly tackle new challenges. The UNCSD mainly discussed the role of green economy in sustainable development and poverty eradication and the institutional framework of sustainable development. The Conference reached consensus on following the principle of "common but differentiated responsibilities" and other basic principles, such as diversification of development patterns, multi-party engagement and consensus. Many countries also put forward suggestions on establishing sustainable development goals, studying and designing new indicators for measuring sustainable development.

In 2015, over 150 heads of state and government met at the UN Headquarters in New York for the UN Sustainable Development Summit, adopting the Transforming Our World: The 2030 Agenda for Sustainable Development. This called on all countries to take action in five critical fields—people, planet, prosperity, peace and partnership—while officially adopting 17 sustainable development goals (Fig. 1.1), which are designed to comprehensively tackle development issues in the three dimensions of society, economy and environment from 2015 to 2030, and move towards a road to sustainable development.

Fig. 1.1 United Nations sustainable development goals

1.3.2 Studies on Ecological Economy

Ecological economy refers to using the theory of ecological economics and systematic engineering approach to change the production and consumption modes, unlock the potential of all available resources, develop economically advanced and ecologically efficient industries, and build an environment with suitable systems, a culture of social harmony and healthy ecology and landscapes, which should be achieved within the carrying capacity of the ecological system.

The concept of ecological economy can be dated back to economists' thoughts about the ultimate carrying capacity of the earth (Qi Hongqian et al. 2016). Kenneth Boulding, an American economist, first proposed the Spaceship Earth theory in the 1960s, arguing that the earth on which human beings live in space is equivalent to a negligible spaceship. Due to economic development, population growth and environmental pollution in this limited and closed system, the entire ecological system will inevitably collapse (Boulding 1966). Therefore, it is only by making ends meet and building an open and circular ecological system with economic means, combining the production of waste with the taking of raw materials within the system, and realizing a dynamic balance between human demands and natural supply, that humankind can avoid a future crisis of resource depletion and environmental deterioration (Qi Hongqian et al. 2016). The major mode of economic and environmental interaction is shown in Fig. 1.2.

Major contributions of the studies on ecological economy are as follows:

Firstly, the concept of ecological economic system was initially raised, defined as a compound system in which the ecological system and economic system are intertwined, interactional and mutually restricted. It is a structural unit of material circulation, energy conversion, value increment and information transmission constituted of technological intermediaries and human labor process (Yan Lidong 1996). Each and every economic process is built on ecological systems and cannot exist independently without these systems (Fu Bojie 2010). This has

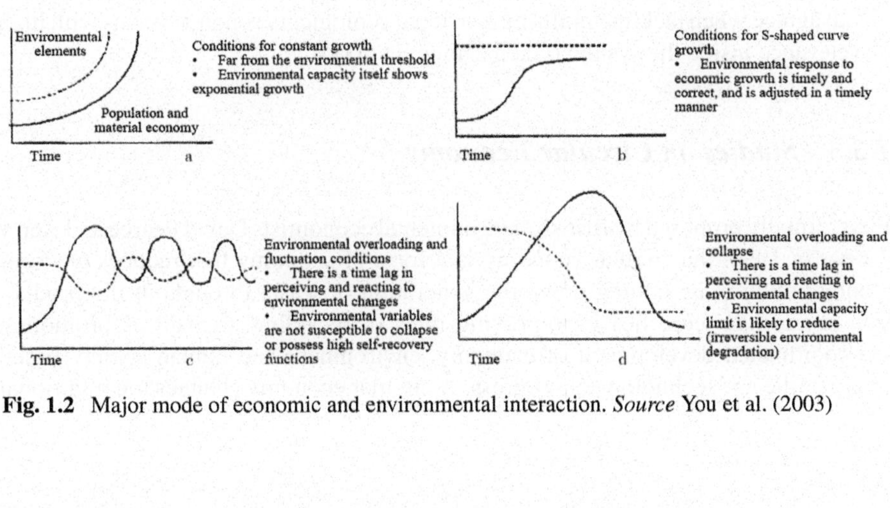

Fig. 1.2 Major mode of economic and environmental interaction. *Source* You et al. (2003)

provided a theoretical support for coordinating the relationship between economy and ecological environment.

Secondly, the concepts of ecological carrying capacity and ecological footprints (EF) were proposed due to the idea that economy should be developed within the scope of ecological carrying capacity. Originating from early studies on environmental carrying capacity (ECC) (Qi Hongqian et al. 2016), this also includes resource carrying capacity and the self-sustaining capacity of the ecological system (Feng Zhiming et al. 2017). Such carrying capacity is mainly manifested by internal adjustment capacity of ecological systems—self-sustaining capacity, and external constraint capacity—press-bearing capacity of the system (Qi Hongqian et al. 2016). EF analysis was first presented by the well-known Canadian scholar William Rees (Rees 1992). EF refers to the supporting and absorption capacity of natural environment within the specific ecological system, which is necessary for humankind's production and consumption activities. It mainly covers two aspects—resource demand and waste absorption. Like the footprints left by human beings on the earth, spatial remnants such as cities and factories formed from the social and economic activities of humankind not only affect, but also depend on the ecological environment. When the theoretical carrying area exceeds the actual carrying level, ecology is in a sustainable surplus state; otherwise, it is in an unsustainable deficit state. Studies of ecological carrying capacity and EF have provided technical means for controlling the environmental benefits of economic activities.

Thirdly, an ecological cost–benefit analysis was conducted. On the basis of evaluating ecological benefits, the analysis also considers the impact of cost factors on the expected net earnings of ecology. The core concept is that the cost and expected earnings of a project are converted to a corresponding monetary value, and then the expected profit of the project is judged by calculating the net monetary income. This can provide a sound basis for whether to implement the project from a quantitative perspective, not only studying the external ecological system with regards to economic benefits, but also analyzing the ecological benefits inside the system. In particular, the ecological cost–benefit analysis enjoys an obvious advantage when tackling multi-dimensional ecological economic issues with high relevance inside the system (Chen Zhongxin 2000).

1.3.3 Studies on Circular Economy

First formally proposed by British environmental economists David Pearce and Kerry Turner in 1990, the circular economy is a form of economy that reduces environmental pollution by cutting down the generation of waste throughout the production process. Its core lies in improving rates of recycling, as well as promoting the coordinated development of economy, environment and human society based on scientific and technological progress. A circular economy changes the traditional economic process of "natural resources—products—wastes" to "natural resources—products—renewable resources", following the principles of reduce, reuse, and recycle (Qu Geping 2001; Chen Demin 2004).

This involves:

Reduce—the use of fewer raw materials and energy inputs to reach the designated production or consumption purpose, controlling the generation of waste through prevention rather than end treatment.

Reuse—extending the time intensity of products and services, using resources as many times or in as many ways as possible, so as to keep them from becoming waste untimely.

Recycle—reducing the final treatment quantity by changing waste to resources again.

1.3.4 Studies on Low-Carbon Economy

As an economic form that emerged to cope with global climate change, the low-carbon economy aims to reduce GHG emissions to a certain level, thereby protecting countries and their peoples from the adverse impacts of global warming, and safeguarding sustainable human settlements across the world (Pan Jiahua et al. 2010). It was first proposed by the British government in the white paper *Our Energy Future—Creating a Low Carbon Economy* released on February 24, 2003, in which the UK said the country would achieve its goal of reducing carbon emissions by about 60% by 2050, compared to levels in 1990 (UK Government 2003). With such core aspects as reduced carbon emissions, increased carbon production efficiency and parallel economic growth, a low-carbon economy is achieved mainly by improving technical efficiency, optimizing energy use mix and industrial structure (Pan Jiahua et al. 2010). In essence, it involves energy efficiency and energy structure (Yan Xin 2020).

The development mode of the low-carbon economy is a green mode, based on low energy consumption, low pollution, low emissions, as well as high effectiveness, high efficiency and high benefits. This mode follows the direction of low-carbon development through energy conservation and emissions reduction, using the method of carbon neutrality technology (Fu Yun et al. 2008). This development mode can be adjusted via market or administrative means. Administrative means work to control GHG emissions through a raft of measures, including enacting laws, levying taxes on polluters, and granting subsidies to enterprises using clean energy and low-carbon technologies. Market means aim to realize the optimized allocation of environmental resources via carbon emissions trading.

On September 22, 2020, President Xi Jinping made a solemn commitment at the General Debate of the 75th Session of the United Nations General Assembly that China would scale up its Intended Nationally Determined Contributions (INDCs) by adopting more vigorous policies and measures, aiming to have CO2 emissions peak before 2030 and to achieve carbon neutrality before 2060. On October 24, 2021, the *Working Guidance of the Communist Party of China Central Committee and the State Council for Carbon Dioxide Peaking and Carbon Neutrality in Full and Faithful Implementation of The New Development Philosophy* was released, stating overall arrangements for China's carbon peaking and carbon neutrality. Decoupling economic development from carbon emissions, and promoting the realization of carbon peaking and carbon neutrality are the key tasks China will face in social and economic development, requiring China to coordinate economic development and environmental protection by seizing the opportunity of carbon peaking and carbon neutrality work.

Chapter 2
Connotation Analysis

2.1 Connotation

The concept of "Lucid Waters and Lush Mountains are Invaluable Assets" is part and parcel of Xi Jinping Thought on Ecological Civilization. While delivering a speech and answering questions from students at Nazarbayev University, Kazakhstan in September 2013, President Xi Jinping systematically and comprehensively expounded on the concept, saying "We want to have not only lucid waters and lush mountains, but also mountains of gold and silver. If we must choose between the two, we would rather have lucid waters and lush mountains than mountains of gold and silver. And in any case, lucid waters and lush mountains are invaluable assets." It can be seen that the concept of "Lucid Waters and Lush Mountains are Invaluable Assets" has three levels of meaning and its connotation can be understood from the following aspects.

2.1.1 Understanding from the Perspective of Dialectics

"Lucid waters and lush mountains" and "mountains of gold and silver" are presented as two sides of a contradiction. It is important to fully recognize their dialectical unification relationship in practice, advancing them in a coordinated way while concentrating on critical points. "We want to have not only lucid waters and lush mountains, but also mountains of gold and silver" shows that we should coordinate environmental protection and economic development, supporting the theory that everything has two aspects in dialectics. Looking at the relationship between environmental protection and economic growth in a one-sided, isolated and static way not only undermines and destroys the ecological environment, but also comes with a heavy price. "If we must choose between the two, we would rather have lucid waters and lush mountains than mountains of gold and silver" shows the focal-point theory of materialistic

H. Zhang et al., *Theory and Practice of "Lucid Waters and Lush Mountains Are Invaluable Assets" in China*, https://doi.org/10.1007/978-981-97-7340-4_2

dialectics. Among them, lucid waters and lush mountains are the focal point and are fundamental. If only one can be chosen from the two, lucid waters and lush mountains is preferred to mountains of gold and silver. "Lucid waters and lush mountains are invaluable assets" fully demonstrates the internal integrity of a contradiction, pointing out that environmental protection underpins economic development, and that lucid waters and lush mountains can be changed to mountains of gold and silver under certain conditions while mountains of gold and silver should support environmental protection. The dialectical unification relationship between lucid waters and lush mountains and mountains of gold and silver has provided an important theoretical support for correctly handling the relationship between economic growth and environmental protection.

2.1.2 Understanding from the Period of Recognition

Recognizing the relationship between "lucid waters and lush mountains" and "mountains of gold and silver" includes three periods. In the first period, lucid waters and lush mountains are blindly exploited in exchange for mountains of gold and silver, without considering or rarely considering the ECC. In the second period, while asking for mountains of gold and silver, lucid waters and lush mountains are protected. The contradiction between economic development and resource scarcity and environmental degradation then becomes apparent, and people realize that environment is the foundation for our survival and development. In the third period, people realize that lucid waters and lush mountains can continuously bring mountains of gold and silver, and that they are in themselves invaluable assets. Ecological advantage can turn into economic advantage. The three periods fully reflect the trajectory of change in the value orientation of development—from economic priority to equal emphasis on economic development and ecological protection, and then to ecological value priority and environmental protection becoming an internal variable of economic development. This is a historical course with the constant improvement and evolvement of the relationship between economic development and environmental protection.

2.1.3 Understanding from the Diversity of Value

Our understanding of the concept of "Lucid Waters and Lush Mountains are Invaluable Assets" should not be confined to the opinion that lucid waters and lush mountains can bring boundless economic wealth. Rather, we should see the ecological value of protecting the environment, such as a healthy and stable environment with steadily improving biodiversity. We should also see social wealth thanks to the increased capabilities of the ecological system, including an improved sense of gain and happiness. Lucid waters and lush mountains do not merely form natural and ecological wealth, but also economic and social wealth. The ecological environment is the most fundamental condition for human survival, and it has no substitute. We simply cannot survive without the ecological environment. Only by better protecting forests, grasslands, wetlands, lakes, oceans and other ecological spaces and resources, can they provide us with more clean air, clean water sources, safe food and other high-quality ecological products, and thus accommodate people's aspirations for a better life (Pan Jiahua 2017).

2.1.4 *Understanding from the Differences in Development Conditions*

The relationship between economic development and environmental protection varies by region and development stage, so it is not feasible to unilaterally pursue the transformation of "lucid waters and lush mountains" to "mountains of gold and silver". Regions with good conditions need to coordinate the relationship between economic development and environmental protection. Protecting the ecological environment means better development, not no development. The better the ecological environment, the stronger the accumulation force of production factors, enabling regions such as these to better promote faster economic and social development, and to realize the transformation of ecological and environmental advantages toward economic and social development advantages. In regions with relatively poor conditions or important ecological functions, such as key ecological functional zones, ecologically vulnerable areas (EVAs) and areas where economic activities surpass the carrying capacity of the ecological environment, a strategic focus on ecological civilization should be maintained, protecting the ecological environment just like protecting eyes, treating the ecological environment just like life, giving greater prominence to environmental protection, and refraining from living beyond one's means, seeking quick success and instant benefits.

2.2 Formation and Development

The concept of "Lucid Waters and Lush Mountains are Invaluable Assets" is a key theoretical concept proposed by President Xi Jinping, based on long-term hands-on experience and a profound understanding of economic development and environmental protection. While working in Hebei, Fujian, Zhejiang and other provinces, President Xi put his thinking into local work, leading both the local government officials and the masses to work together through concrete actions, changing the previous old road of economic development and letting the green development concept enjoy popular support.

During his work in Hebei, he was in charge of formulating the *General Plan for Economic, Technological and Social Development of Zhengding County*, which emphasized that "protecting the environment, abating pollution, managing, developing and exploiting resources, and maintaining ecological balance are the key tasks of modernization, and also the eager requests of people's production and life". The document particularly stressed that "we would rather not be affluent than allow pollutants. We should strictly prevent pollution from moving to the countryside and beyond".

During his tenure in Fujian, President Xi proposed placing forestry revitalization in the strategic position of economic development. This was an important step in spurring on economic development in eastern Fujian, and encouraged the launch of "green projects". He also proposed making the best use of local resources such as mountains and seas, and prospectively put forward the strategic concept of building Fujian into an "ecology-priority province", initiating the largest-ever ecological protection project in Fujian.

Over the course of his work in Zhejiang, he further reflected on the relationship between economic development and environmental protection, producing many important viewpoints and statements. As a result of this, he officially put forward the concept of "Lucid Waters and Lush Mountains are Invaluable Assets". In November 2002, Xi Jinping, then secretary of the Party Committee of Zhejiang Province, indicated when leading a team on a survey in Lishui that "ecological advantage should not be discarded. In the post-industrialization era, this [losing that advantage] is the biggest regret for those industrialized areas and areas that failed to take ecological protection seriously. Never sacrifice the environment for a little economic benefit."

On July 11, 2003, he delivered an inspiring speech at the mobilization meeting to build Zhejiang into an ecology-priority province. His speech stated that "building Zhejiang into an ecology-priority province is a strategic task that is key to the overall situation and of long-term importance, and is a grand systematic project. Minimum cost of resources and environment should be paid for maximum economic and social development, and minimum social and economic cost should be used to protect resources and the environment. We should not sacrifice the environment for development, nor should we give up development simply for the sake of protection."

On August 15, 2005, Xi Jinping, then secretary of the Party Committee of Zhejiang Province, commended villagers of Yucun, Anji, Zhejiang Province for closing mines as a "wise step". He officially put forward the concept of "Lucid Waters and Lush Mountains are Invaluable Assets", encouraging the village of Yucun to pursue green development. Nine days after his survey at the village, he published a short piece of commentary entitled "*Lucid Waters and Lush Mountains are also Invaluable Assets*" in the "*Zhejiang, China: A New Vision for Development*" column of *Zhejiang Daily*, saying "simply put, we pursue harmony between humankind and nature, and between economy and society. This means that we want not only lucid waters and lush mountains, but also mountains of gold and silver". He then put forward the analysis that "Zhejiang Province enjoys a geographical makeup of 'more than 70% mountainous areas, less than 10% water areas, and about 20% flatlands'. Many areas have the picturesque scenery that he described as 'the lucid water of the Yangtze

River twists and turns and continues to flow forward. Lush mountains on both sides of the river make way for riverway along the ships', and possess sound ecological advantages. If these ecological and environmental advantages can be turned into advantages in ecological agriculture, ecological industry, ecological tourism and other ecological economy, then lucid waters and lush mountains will indeed become mountains of gold and silver." Finally, he elaborated on the dialectical relationship between lucid waters and lush mountains and mountains of gold and silver, saying that "lucid waters and lush mountains can bring mountains of gold and silver, but mountains of gold and silver cannot buy lucid waters and lush mountains. They are contradictory, yet dialectically unified". As you cannot have the two things at the same time, we must be good at making choices, learn to abandon things when necessary, when to do something and when to leave some things undone. Conditions can be created selectively, enabling lucid waters and lush mountains to constantly bring mountains of gold and silver.

On September 7, 2013, when delivering a speech and answering questions from students of Nazarbayev University, Kazakhstan, President Xi Jinping said that "We want to have not only lucid waters and lush mountains, but also mountains of gold and silver. If we must choose between the two, we would rather have lucid waters and lush mountains than mountains of gold and silver. And in any case, lucid waters and lush mountains are invaluable assets."

In 2015, the concept of "Lucid Waters and Lush Mountains are Invaluable Assets" was incorporated into the *Guidelines on Accelerating Ecological Civilization*, marking its integration into the central document. In October 2017, the concept of "Lucid Waters and Lush Mountains are Invaluable Assets" was also included in the report of the 19th CPC National Congress, gradually becoming an important guidance that directs China's environmental protection and social and economic development, and a guiding ideology to boost China's ecological civilization.

Since then, whether at important conferences or during field trips and visits, President Xi Jinping has touched upon the principle of "Lucid Waters and Lush Mountains are Invaluable Assets" on multiple occasions. This concept of "Lucid Waters and Lush Mountains are Invaluable Assets" has significantly affected and changed China's development philosophy, become the fundamental tone and one of the basic ideas of Xi Jinping Thought on Ecological Civilization, and has developed and formed a set of systematic, scientific and complete theories.

2.3 Significance

In today's world, climate change, environmental pollution and decreasing biodiversity have become key factors threatening human survival and development. Various countries regard "achieving harmony between man and nature" as a key goal in responding to ecological environmental issues around the world. The United Nations launched the action plan "UN Decade on Ecosystem Restoration", calling upon all stakeholders to take actions to protect and restore the ecological system in the ten years of ecosystem restoration from 2021 to 2030, help restore the health and vitality of the ecological system, and to support and promote the achievement of sustainable development goals (Strassburg et al. 2020).

Xi Jinping Thought on Ecological Civilization and the concept of "Lucid Waters and Lush Mountains are Invaluable Assets" are breakthroughs made in the study of the relationship between economic development and environmental protection, and their theoretical results have been widely practiced and verified, enjoying popular support and helping gain outstanding achievements in China. They have played an important guiding role in addressing major ecological environmental issues faced by the international community, and the significance in the era is shown as follows.

2.3.1 The Concept of "Lucid Waters and Lush Mountains Are Invaluable Assets" Embodies a Major Leap in Epistemology

President Xi Jinping stated that lucid waters and lush mountains are not separate from invaluable assets; to protect the environment is to protect productivity, and to improve the environment is to develop productivity (Xi Jinping 2019). The viewpoints that pits environmental protection and economic development against each other is the root cause of air pollution, water environment pollution, land desertification and sand disaster, water and soil loss, biodiversity destruction and other ecological environmental issues. The concept of "Lucid Waters and Lush Mountains are Invaluable Assets" clearly points out that environmental protection and economic development are dialectically unified, that there is a path to coordination and mutual benefit for both of them, and that lucid waters and lush mountains are invaluable assets.

2.3.2 The Concept of "Lucid Waters and Lush Mountains Are Invaluable Assets" Reflects an Important Innovation of Value Theory

President Xi Jinping noted that natural ecology has value, and protecting nature is a process that increases natural value and natural capital. To change this to economic value and material capital, and to coordinate development and protection at a higher level, we must constantly explore the path of transforming lucid waters and lush mountains to invaluable assets, establish and improve the value realization mechanism of ecological products, enable lucid waters and lush mountains to continuously leverage ecological, economic and social benefits, and to realize the outcome of "gaining invaluable assets while maintaining lush mountains".

2.3.3 The Concept of "Lucid Waters and Lush Mountains Are Invaluable Assets" Provides an Important Breakthrough in Methodology

President Xi Jinping indicated that we should correctly handle the relationship between economic development and environmental protection, and not sacrifice the environment for temporary economic growth. Economic downward pressure is currently mounting, and global risks and challenges are evidently growing. Under the influence of multiple factors including intertwined structural, institutional and periodic issues, and the severe impact of the COVID-19 pandemic, there may be signs of sacrificing the ecological environment for economic growth. The concept of "Lucid Waters and Lush Mountains are Invaluable Assets" has provided an important principle for solving current ecological environmental issues, and coordinating a win–win situation for environmental protection and economic and social development while guaranteeing an unchanged strategic focus of maintaining the direction and intensity of environmental protection.

2.3.4 The Concept of "Lucid Waters and Lush Mountains Are Invaluable Assets" Provides Chinese Approach and Chinese Wisdom for Global "Green Governance"

Green development and sustainable development are key concepts of global governance, forming a goal and direction that countries across the world pursue together. Despite positive progress in sustainable development around the world, the UN's latest evaluation report shows that 17 goals in the *2030 Agenda for Sustainable Development* still face daunting challenges (Sachs et al. 2024). The concepts of "community with a shared future for mankind", "harmony between man and nature" and "a win–win situation for environmental protection and economic growth", and other ideas contained in the concept of "Lucid Waters and Lush Mountains are Invaluable Assets" boast great theoretical, practical and universal property, offer new, feasible ways and approaches for the world's environmental protection and sustainable development, gain extensive recognition of the international community, and can surely magnify the wellbeing of humanity.

Chapter 3
Practice Progress

The 19th CPC National Congress incorporated the idea that "we should fully understand that lucid waters and lush mountains are invaluable assets" into the newly-amended *Constitution of the Communist Party of China*, and integrated the statement that "we must establish and practice the concept of lucid waters and lush mountains are invaluable assets" into Xi Jinping Thought on Socialism with Chinese Characteristics for a New Era as an important aspect. The concept of "Lucid Waters and Lush Mountains are Invaluable Assets" has become one of the important governing ideas of the Party, and has begun to be comprehensively integrated into China's top-level design.

In April 2018, President Xi Jinping mentioned at the Seminar on Deeply Promoting the Development of the Yangtze River Economic Belt that "it is necessary to actively explore and popularize the paths of transforming lucid waters and lush mountains to invaluable assets, choose well-developed regions to pilot the value realization mechanism of ecological products, and explore a sustainable value realization path of ecological products that is led by the government, participated in by enterprises and people from all walks of life, and features market-based operation".

In May 2018, President Xi Jinping said at the National Conference on Ecological and Environmental Protection that we should follow the concept of lucid waters and lush mountains are invaluable assets, implement the philosophy of "innovative, coordinated, green, open and shared development", rapidly establish spatial pattern, industrial structures, production modes and lifestyles that conserve resources and protect the environment, leaving time and space for the natural ecology to be rehabilitated.

In June 2018, the CPC Central Committee and the State Council released the Opinions on Comprehensively Enhancing Ecological and Environmental Protection and Resolutely Winning the Tough Battle for Prevention and Control of Pollution, clarifying that we should "promote the creation of ecological civilization demonstration, and construct 'Lucid Waters and Lush Mountains are Invaluable Assets'

H. Zhang et al., *Theory and Practice of "Lucid Waters and Lush Mountains Are Invaluable Assets" in China*, https://doi.org/10.1007/978-981-97-7340-4_3

Practice and Innovation Bases", thereby increasing the speed of the formulation of an ecological environment governance system.

In November 2020, President Xi Jinping emphasized at the Seminar on Comprehensively Promoting the Development of the Yangtze River Economic Belt that the establishment of value realization mechanism of ecological products should be accelerated, whereby protection and restoration of the ecological environment shall be reasonably rewarded, while those responsible for ecological environment destruction shall be suitably penalized.

In April 2021, the General Office of the CPC Central Committee and the General Office of the State Council released the *Opinions on Establishing and Improving the Value Realization Mechanism of Ecological Products*, implementing the concept of "Lucid Waters and Lush Mountains are Invaluable Assets" at the levels of institutional arrangement and practical operation, and requiring that under the premise of strictly protecting the ecological environment, we should encourage the adoption of diversified modes and paths, scientifically and rationally advancing the value realization of ecological products.

In October 2021, the CPC Central Committee and the State Council released the *Outline for the Ecological Protection and High-quality Development of the Yellow River Basin*. This document set out requirements for improving the value realization mechanism of ecological products in the upstream reaches of the Yellow River, starting from major tributaries, rivers and lakes, expediting pilot projects and experiments, and improving the compensation mechanism for ecological protection of the Yellow River Basin with vertical and horizontal coordination, supplemented by marketization.

3.1 Practices and Explorations

Under the deployment of the CPC Central Committee and the State Council, relevant ministries and commissions of China have advanced ecological civilization in a steady and orderly manner, as per the assignment of responsibility by deeply fulfilling the concept of "Lucid Waters and Lush Mountains are Invaluable Assets". They have conducted "Lucid Waters and Lush Mountains are Invaluable Assets" pilot demonstrations at national and provincial levels, rolled out a series of guiding opinions, management regulations, technical guidelines and other policy documents and standard specifications, with a view to guiding and leading various regions to actively carry out practices and explorations in this field.

In 2017, the MEE launched the construction of "Lucid Waters and Lush Mountains are Invaluable Assets" practice and innovation bases and formulated and released the *Regulation on the Administration of the Establishment of "Lucid Waters and Lush Mountains are Invaluable Assets" Practice and Innovation Bases (For Trial Implementation)*, among other policy documents. As the platform and carrier for fulfilling Xi Jinping Thought on Ecological Civilization and the concept of "Lucid

Waters and Lush Mountains are Invaluable Assets", "Lucid Waters and Lush Mountains are Invaluable Assets" practice and innovation bases are designed to innovate and explore institutional and implementation practices of transforming "lucid waters and lush mountains" to "invaluable assets", summarizing and popularizing typical experiences, and encouraging regional green development. 136 "Lucid Waters and Lush Mountains are Invaluable Assets" practice and innovation bases have currently been named, and a group of practice samples fulfilling the concept of "Lucid Waters and Lush Mountains are Invaluable Assets" have been cultivated and created across China. The *"Lucid Waters and Lush Mountains are Invaluable Assets" Practice Modes and Typical Cases (Batch 1)* was published at the 2021 Eco Forum Global Guiyang, summarizing and forming four typical transformation paths of "Safeguarding Lucid Waters and Turning Them into Invaluable Assets", "Increasing Lucid Waters and Invaluable Assets", "Turning Lucid Waters into Invaluable Assets", and "Green Capital", as well as eight modes of "ecological restoration, ecological agriculture, ecological tourism, ecological industry, 'ecology+' compound industry, ecological market, ecological finance, and ecological compensation". It serves as a sound reference and guide for various regions to practice "Lucid Waters and Lush Mountains are Invaluable Assets". At the Fifteenth Meeting of the Parties to the Convention on Biological Diversity (COP-15), an English edition of *"Lucid Waters and Lush Mountains are Invaluable Assets" Practice Modes and Typical Cases* (Batch 1) was released to the world, vividly demonstrating China's remarkable achievements in ecological civilization under the guidance of Xi Jinping Thought on Ecological Civilization and providing Chinese Wisdom and Chinese Approach for global ecological civilization.

Column 1 "Lucid Waters and Lush Mountains are Invaluable Assets" Practice and Innovation Bases

---

"Lucid Waters and Lush Mountains are Invaluable Assets" practice and innovation bases are the practical platforms and carriers for fulfilling Xi Jinping Thought on Ecological Civilization and the concept of "Lucid Waters and Lush Mountains are Invaluable Assets", designed to innovate and explore the institutional and implementation practice of transforming "lucid waters and lush mountains" to "invaluable assets", to summarize and popularize typical experiences

Application conditions:

(I) Sound ecological environment, and solid foundation for environmental protection work;
(II) Evident results in transforming "lucid waters and lush mountains" to "invaluable assets", with typical transformation cases involving townships, villages or small river basins;
(III) Effectively promoting the transformation of "lucid waters and lush mountains" to "invaluable assets" through systems and mechanisms; and
(IV) No major issues found in the central supervisions on environmental protection and various special supervisions in the past three years, and no major ecological environmental destruction events

---

Implementation:

The MEE has currently shortlisted and named 136 regional "Lucid Waters and Lush Mountains are Invaluable Assets" practice and innovation bases in five batches

The first batch of 13 regions were named in September 2017;
The second batch of 16 regions were named in December 2018;
The third batch of 23 regions were named in November 2019;
The fourth batch of 35 regions were named in October 2020;
The fifth batch of 49 regions were named in October 2021

The National Development and Reform Commission (NDRC) has deeply implemented the Opinions on Establishing and Improving the Value Realization Mechanism of Ecological Products, organized and conducted the pilot demonstrations of the value realization mechanism of ecological products, and intensified innovations in establishing accounting systems, formulating accounting specifications and promoting the application of accounting results in light of different value realization paths of ecological products, effectively tackling "difficulties in measurement". The NDRC has also enhanced exploration on advancing precise connections of supply and demand, expanding value realization modes, promoting value increments, and advancing equity trading of ecological resources, effectively tackling "difficulties in trading". It has held in-depth studies on improving protection and compensation mechanisms, perfecting damage compensation systems, and on establishing evaluation mechanisms, effectively tackling "difficulties in realization". It has also worked to guide financing institutions to offer green credit, innovate green finance and open new fields of green finance, effectively tackling "difficulties in mortgage". It has also propelled the Yangtze River Economic Belt to conduct green development "5 + 2" pilot demonstrations, initially forming a series of typical experience and practices, such as "protecting lucid waters as the foundation, building up green development; increasing lucid waters through industries, promoting green transition and upgrading; facilitating lucid waters through science and technology, cultivating and developing new growth drivers; turning lucid waters to invaluable assets, exploring new paths of transforming 'lucid waters and lush mountains' to 'invaluable assets'; transforming lucid waters, and enhancing institutional innovation".

Column 2 Opinions on Establishing and Improving the Value Realization Mechanism of Ecological Products
In 2021, the General Office of the CPC Central Committee and the General Office of the State Council issued the Opinions on Establishing and Improving the Value Realization Mechanism of Ecological Products. This was the first framework document implementing the concept of "Lucid Waters and Lush Mountains are Invaluable Assets" at the levels of institutional arrangement and practical operation in the 15 years since President Xi Jinping put forward the concept of "Lucid Waters and Lush Mountains are Invaluable Assets"
Two phased goals: By 2025, China will preliminarily establish an institutional framework for the value realization of ecological products, formulate a scientific system for accounting the value of ecological products, gradually improve policies and systems for ecological protection compensation and environmental damage compensation, establish a government assessment mechanism for the value realization of ecological products, effectively tackle issues of ecological products, including "difficulties in measurement, mortgage, trading and realization", establish an interest-orientation mechanism for environmental protection, and build capacity in transforming ecological advantages to economic advantages By 2035, China will comprehensively establish an excellent value realization mechanism of ecological products, comprehensively formulate a new mode of ecological civilization with Chinese characteristics, and extensively form green ways of production and life, thereby providing strong support for realizing the goal of building a beautiful China

(continued)

(continued)

Establishing and improving six mechanisms:

(I) Establishing survey and supervision mechanisms of ecological products: advancing the ownership registration of natural resources and conducting an information census on ecological products,

(II) Establishing a value evaluation mechanism of ecological products: establishing a value evaluation system of ecological products, formulating value accounting specification of ecological products, and pushing forward the application of value accounting results of ecological products;

(III) Improving operation and development mechanisms of ecological products: driving forward the precise connection of supply and demands of ecological products, expanding the value realization modes of ecological products, promoting the value increment of ecological products, and advancing the equity trading of ecological resources;

(IV) Improving the protection and compensation mechanism of ecological products: perfecting the vertical ecological protection compensation system, establishing a horizontal ecological protection compensation mechanism, and improving the environmental damage compensation system;

(V) Improving the guarantee mechanism for the value realization of ecological products: establishing a value assessment mechanism of ecological products, establishing an interest-orientation mechanism for environmental protection, intensifying support for green finance; and

(VI) Establishing an advancing mechanism for the value realization of ecological products: enhancing organization and leadership, advancing pilot demonstrations, intensifying intelligent support, promoting and encouraging implementation

In 2020, the Ministry of Natural Resources of China (MNR) successively issued a total of 21 typical cases in China and abroad regarding the value realization of ecological products, including the Typical Cases of Value Realization of Ecological Products (Batch 1) and the Typical Cases of Value Realization of Ecological Products (Batch 2), serving as guidance for learning and practice in various regions. In 2021, the MNR issued the Working Instruction for Piloting the Value Realization Mechanism of Ecological Products in the Field of Natural Resources, requiring guiding and encouraging regions to boldly explore based on local conditions under the main theme of promoting the value realization of ecological products, work to build theoretical systems, technical systems and policy systems for the value realization of ecological products in the field of natural resources, and explore replicable and propagable paths for the value realization of ecological products.

Column 3 Typical Cases of Value Realization of Ecological Products

In April 2020, the General Office of the MNR issued the *Typical Cases of Value Realization of Ecological Products (Batch 1)*, summing up four types of major practices involving the value realization of ecological products, including ecological resource indicators and equity trading, ecological restoration and value enhancement, industrialized ecological operation and ecological compensation. Five critical links promoting the value realization of ecological products were proposed—these included following planning guidance and scientific and reasonable arrangements, controlling and creating demands and cultivating trading markets, clarifying equity and promoting equity transfer, developing ecological agriculture and stimulating market vitality, formulating supporting policies and realizing value spillover. In the *Batch 1*, 11 typical cases were recommended, including Wuyuanwan of Xiamen, Fujian Province, "Forest Eco-bank" of Nanping, Fujian Province, and expansion of the ecological function of land quota, Chongqing Municipality

In October 2020, the General Office of the MNR issued the *Typical Cases of Value Realization of Ecological Products (Batch 2)*, recommending ten typical cases, including the development of "ecological agriculture, culture and tourism" in Jinting Town, Suzhou, Jiangsu Province, "water-enabled economy" of Guangze County, Nanping, Fujian Province, and development of ecological industries in Xichuan County, Henan Province

The National Forestry and Grassland Administration of China (NFGA) has actively fulfilled the concept of "Lucid Waters and Lush Mountains are Invaluable Assets", pushed forward ecological restoration and protection, created lucid waters and lush mountains, intensified the restoration and protection of vegetation cover, and ensured mountains are lusher, waters are clearer and the environment is more beautiful. It has advanced efforts to benefit people via ecology and promoted the development of industries, developed invaluable assets, and cultivated new industries, new formats and new economic growth drivers, turning resources into assets and capital and turning lucid waters and lush mountains as well as frozen and snow-covered land into invaluable assets. The NFGA has established a formation mechanism of "Lucid Waters and Lush Mountains are Invaluable Assets", improved forestry operation and management systems and mechanisms, forestry enhancement and beneficial policy systems, and co-building and sharing participation mechanisms. It has popularized advanced models and ecological culture, publicized and promoted typical models that fulfill the concept of "Lucid Waters and Lush Mountains are Invaluable Assets", and better played a guiding and leading role in typical models. It has continuously boosted the development of forest cities and villages. The *2020 Bulletin of Land Greening in China* shows that 441 cities engaged in the development of national forest cities in 2020.

Column 4 "Internet + Nationwide Voluntary Tree-planting" Pilot Projects
Nationwide voluntary tree-planting represents a key move in speeding up land greening under the new circumstances, and is also a specific practice in fulfilling the concept of "Lucid Waters and Lush Mountains are Invaluable Assets" proposed by President Xi Jinping
In 1981, the fourth session of the Fifth National People's Congress made the *Resolution on Launching a Nationwide Voluntary Tree-planting Campaign*, laying down the tree-planting obligation of citizens through legislation. Since the 18th CPC National Congress, President Xi Jinping has attended voluntary tree-planting activities each year, and made important instructions on tree-planting and afforestation for many times. In 2017, the National Greening Commission issued the *Measures for the Administration of the Forms in Fulfilling the Nationwide Voluntary Tree-planting Obligation (For Trial Implementation)*, enriching and expanding the forms of fulfilling obligation to over 50 in eight types, including afforestation, tending, management and protection, natural protection, and claiming the planting of trees, and also specifying meticulous conversion standards
On September 1, 2017, the National Greening Commission gave an official statement on four provinces and municipalities, including Beijing, as the first batch of national pilot provinces and municipalities of "Internet + Nationwide Voluntary Tree-planting". By 2020, the number of such pilot provinces and municipalities in China had reached 15
Over the years, the National Greening Commission has innovatively released a group of online voluntary tree-planting projects, including "I Send Support to Ancient Trees and Famous Trees", and "I Plant Bamboo for Giant Pandas". It has also established the first batch of 26 national "Internet + Nationwide Voluntary Tree-planting" bases, making "planting trees as you wish, anywhere and anytime" a reality

Under the guidance of the "double eight strategy" and the concept of "Lucid Waters and Lush Mountains are Invaluable Assets", Zhejiang Province has constantly explored a group of projects, including developing Zhejiang into an ecology-priority province, "Thousand Village Demonstration and Ten Thousand Village Renovation" project, the first trans-provincial watershed ecological compensation mechanism in China, and "river chief system". It became the first ecology-priority province in

China in 2020, forming a group of replicable and propagable "Zhejiang experiences". In 2019, the Committee for Comprehensively Deepening Reform of the Zhejiang Provincial Party Committee issued the *Overall Plan of Comprehensive Reform and Innovation Pilot Zone of Zhejiang (Anji) for Fulfilling the Concept of "Lucid Waters and Lush Mountains are Invaluable Assets" in the New Era*, endowing Anji with the new mission of becoming a forerunner to get through the new channel for transformation gradually, and providing a Zhejiang approach for building a beautiful China. As the first pilot city for the value realization mechanism of ecological products in China, Lishui has made constant reforms and innovations by centering on the Gross Ecosystem Product (GEP), realizing double growth in GEP and GDP.

Column 5 "Double Eight Strategy": Eight Future-oriented Measures

The "Double eight strategy" includes eight future-oriented measures proposed by Xi Jinping, the then Secretary of the Party Committee of Zhejiang Province, at the Fourth Plenary Session of the 11th Party Committee of Zhejiang Province held in July 2003, i.e., further leveraging strengths in eight areas and advancing measures in eight areas

The contents of the strategy are as follows:

First—the institutional strength of Zhejiang should be leveraged, strongly promoting the common development of economy with diverse forms of ownership, dominated by public ownership, and constantly improving the socialist market economy system

Second—the regional advantage of Zhejiang should be leveraged, actively connecting with Shanghai and participate in cooperation and communication in the Yangtze River Delta area, and improving internal and external opening-up

Third—the advantage of bulk characteristic industries of Zhejiang should be leveraged, speeding up the base construction of advanced manufacturing, and embarking on a new road to industrialization

Fourth—Zhejiang's strength must be leveraged through coordinated urban–rural development, and urban–rural integration must be sped up

Fifth—Zhejiang's ecological advantage must be leveraged, building Zhejiang into an ecology-priority province and building a "Green Zhejiang"

Sixth—the strength of Zhejiang in mountain and ocean resources should be leveraged, vigorously developing an ocean economy, promoting the significant development of underdeveloped areas, and making the development of ocean economy and underdeveloped areas new economic growth points of Zhejiang

Seventh—the environmental strength of Zhejiang should be leveraged, actively advancing the development of key projects, with "five major CNY10 billion projects" as the major contents, and earnestly enhancing the development of judicial system, credit and effectiveness of organizations

Eight—the cultural strength of Zhejiang should be leveraged, actively advancing the strategy of developing Zhejiang through science, education and talents, and rapidly building Zhejiang into a strong cultural province

Under the guidance of the "double eight strategy", Zhejiang Province has achieved "three leapfrog changes" in recent years from a large economic province to a strong economic province, from internal and external opening-up to deep integration into the world, and from well-off in an all-round way to well-off at a high level

Fujian Province has issued the *Rules on Promoting Ecological Civilization of Fujian Province*, marking the fact that the concept of "Lucid Waters and Lush Mountains are Invaluable Assets" is incorporated throughout the legislation of Fujian Province for the first time, and becoming an operable and enforceable stipulation. As the first national ecological civilization pilot zone in China, Fujian Province has taken the lead in establishing a target-oriented system in which Party and government leaders take responsibility for environmental protection, and comprehensively advanced environmental pollution prevention and control, construction and management of natural reserves, and afforestation. In 2017, the province issued the *Pilot*

Program for the Redemption of Commercial Forest in Important Ecological Zones and Other Reforms of Fujian Province, and Nanping, as the pilot city, has comprehensively advanced the redemption of commercial forest in important ecological zones, taking the lead in establishing a "Forest Eco-bank" in China, allowing lush mountains to enter the "bank", making forest workers "depositors", and enabling the transformation of "lucid waters and lush mountains" to "invaluable assets". In the same year, Fujian Province issued the *Pilot Program for Forestry Carbon Sink Trading of Fujian Province* and the *Implementation Plan for the Establishment of Green Financial System of Fujian Province*, while working to build Fujian into a province with "vigorous mechanism, superior industries, wealthy residents and beautiful ecology".

Column 6 Pilot of "Forest Eco-bank"

In order to effectively tackle difficulties in the value realization of ecological resources, Nanping, Fujian Province, chose Shunchang County (which enjoys abundant forestry resources yet has high decentralization) to initially pilot its "Forest Eco-bank" in 2018. Lvchang Forestry Resource Operation Co., Ltd. controlled by the state-owned forest station of the county was organized as the operator of "Forest Eco-bank", and has two centers and three companies under it which specialize in forest right evaluation, operation and financial services

"Forest Eco-bank" is not a financial institution, nor is it a financing platform. Instead, it is an operation system composed of the government, natural resource operation and management companies, villagers and cooperatives, and social capital providers. Based on the "decentralized input and integrated output" mode of commercial banks, it has collected, stored, integrated and optimized fragmented resources, and produced high-quality "asset packages", entrusting professional and influential industry operators to conduct management. It has also gotten through the channel for turning resources to assets and turning assets to capital by introducing social capital, exploring a value realization path of ecological products that turns the advantage of ecological resources to economic development advantage

Main measures:

(1) The pilot project has established the mechanism of "forest right deposit receipt". Using policy-based lending at the China Development Bank, a "forest right deposit receipt" business has been opened at "Forest Eco-bank", where forest workers can directly "realize" forest resources through forest right mortgage loans, forest right transfers and other modes
(2) It has served the development of the real economy. It has provided heavy asset services for large-scale wood and bamboo processing enterprises, and the "Forest Eco-bank" has been introduced to construct workshops. Enterprises can start production after buying equipment, which alleviates enterprises' investment burden, and significantly shortens the construction period of projects
(3) It has conducted the development of single plant forest carbon sink. This involves forest resources being "deposited" by forest workers, i.e., a single tree has been taken as the unit, and carbon sink methodology has been used to measure the fixed carbon content of each plant, with the price being fixed after evaluation based on the market, which can be sold through online platforms

3.2 Key Results

The concept of "Lucid Waters and Lush Mountains are Invaluable Assets" has become the consensus and action of the entire Party and society. Relevant ministries and commissions and provinces, autonomous regions and municipalities in China have actively fulfilled the concept of "Lucid Waters and Lush Mountains are Invaluable Assets", deeply conducting practice and explorations, and scoring remarkable results in the protection of lucid waters and lush mountains, transformation of invaluable assets, reform and innovation of systems and mechanisms, and pilot demonstrations. In particular, the construction of "Lucid Waters and Lush Mountains are Invaluable Assets" (hereinafter referred to as "Liangshan") practice and innovation bases organized and conducted by the MEE has become an important platform for fulfilling the concept of "Lucid Waters and Lush Mountains are Invaluable Assets", advancing green development and building a beautiful China, and an important carrier for exploring the value realization of ecological products.

Firstly, environmental protection has gained outstanding outcomes, contributing to sound ecological environmental quality, and consolidating the foundation and basis for the transformation of "Lucid Waters and Lush Mountains are Invaluable Assets". Environmental protection has been taken as a condition and requirement for constructing "Lucid Waters and Lush Mountains are Invaluable Assets" practice and innovation bases, promoting the generation of a group of samples of beautiful China with sound ecological environment and prominent ecological and resource advantages. The "Liangshan" practice and innovation bases have done a good job in fighting a tough battle for the prevention and control of pollution, while their air quality and water environment quality rank top among China's provinces, autonomous regions and municipalities. Moreover, these bases eliminated inferior V-type water bodies and black and odorous water bodies, and achieved the goals and missions involving the number of days with good air quality, $PM_{2.5}$, and water bodies which reach or are higher than III-type. The integrated protection, systematic restoration and comprehensive governance of mountains, waters, forests, farmland, lakes, grasslands and sand have been advanced in a coordinated way, with the pattern of national land development and protection constantly optimized, and the quality and stability of the ecological system seeing a constant improvement. Moreover, the capacity in supplying ecological products has been constantly elevated, ecological indicators have all exceeded the level of excellence, a green economy has been restored and increased, ecological assets have achieved value maintenance, and the "spillover" of the value of ecological products has been continuously enhanced.

Secondly, the value realization of ecological products has been promoted effectively, and the transformation path of "Lucid Waters and Lush Mountains are Invaluable Assets" has been opened up, promoting an evident improvement in local green development. The "Liangshan" practice and innovation bases have developed characteristic ecological agriculture, ecological tourism and ecological industry by focusing on their respective features of natural environment and advantages of ecological resources, constantly extending the upstream and downstream industry chain, attaching the value of ecological products to the value of agricultural products, industrial products and service products, and promoting the value realization of ecological products, thus constantly improving the per capita disposable income and elevating people's wellbeing. They have explored the high-quality development road of "ecological industrialization and industrial ecologicalization" based on local conditions, continuously harnessing the synergy of pollution reduction and carbon reduction, and ceaselessly cut down energy consumption, water consumption, and carbon dioxide emissions per unit of GDP. As a result, the regional green development level has seen a constant increase, producing a group of green development demonstration samples which feature constantly reduced resources and energy consumption, ever-increasing green and low-carbon levels, and gradually optimized and upgraded industries.

Thirdly, institutional innovation has been moved forward, ensuring that lucid waters and lush mountains can continuously bring invaluable assets and providing a long-term guarantee for building ecological civilization and building a beautiful China. Institutional reform has been regarded as an innovative expression of "Lucid Waters and Lush Mountains are Invaluable Assets" practice and innovation bases, guiding various regions to conduct rewarding practice in following and improving ecological civilization systems while innovating and exploring market-based operation and market trading systems for the value realization of ecological products. By setting up public-welfare ecological posts and conducting ecological procurement, they have expanded ecological employment channels, made innovation in the trading system of energy usage rights, carbon emission rights, pollution rights and water rights as well as green finance system, and added to the ways and means of benefiting people through ecology. They have also reformed the market of ecological products, constantly developed ecological financial products and guided and stimulated stakeholders to carry out the market-based trading of ecological products, so as to ensure the long-term transformation of "Lucid Waters and Lush Mountains are Invaluable Assets".

Fourthly, the practice of constructing "Lucid Waters and Lush Mountains are Invaluable Assets" has been advanced thoroughly, formulating a group of vivid practice samples which serve as typical benchmarks. The "Liangshan" practice and innovation bases cover different natural resources including mountainous areas, plains, forest areas and islands, with different regional conditions in the eastern, central and western areas, and have preliminarily generated a group of typical, demonstrative and propagable practice cases and modes concerning the transformation of "Lucid Waters and Lush Mountains are Invaluable Assets". These showcase the scenes of "making the best use of local resources such as mountains and oceans", "better protecting lucid waters and lush mountains", "letting people in the locality get 'rich' based on local waters and mountains", "developing immense forests from a single tree", and "following through on the mission for ultimate success in each session of leadership". They have shaped benchmarks and samples of ecological civilization, standing for different development stages, industrial development modes and functional orientation of different groups in China, providing referential experience and practice for nationwide practice of "Lucid Waters and Lush Mountains are Invaluable Assets".

3.3 Practice Modes

"Lucid Waters and Lush Mountains are Invaluable Assets" practice modes can be summarized as four typical modes, including ecological restoration, ecological industry, ecological market and green finance, by comprehensively considering natural environment, regional features, ecological functions and resource endowment, construction progress and effects, representative and the propagable nature of experience and modes, all in light of the typical transformation paths of "Lucid Waters and Lush Mountains are Invaluable Assets".

3.3.1 Ecological Restoration Mode

The main tasks and measures of ecological restoration mode include improving ecological environmental quality, elevating ecological assets and increasing the value of ecological assets. By continuously conducting ecological protection and restoration and ecological construction projects to increase and replenish ecological products, this mode aims to constantly consolidate the ecological foundation of green and sustainable development, strengthen the "eco-bank" of economic and social development, promote the constant value increment, accumulation and realization of ecological assets and green capital, and turn the natural and ecological wealth of lucid waters and lush mountains into economic and social wealth, thus achieving the coordinated advance of ecological civilization, poverty elimination, rural revitalization and green development. It is mainly applicable to areas with poor ecological environmental foundation or EVAs. For example, by accumulating and putting ecological assets into use with a solid ecological background, the Saihanba Forest Farm (Hebei Province), Youyu County (Shanxi Province), the Elion Ecological Demonstrative Area (Kubuqi, Inner Mongolia Autonomous Region), Huaxiacheng (Weihai, Shandong Province), Luanchuan County (Henan Province), the Babusha Forest Farm (Gulang County, Gansu Province) and other regions have deeply demonstrated the profound historical view of "when ecology prospers, civilization prospers".

3.3.2 Ecological Industry Mode

This mode can be divided into ecological agriculture, ecological industry, ecological tourism and "ecology+" compound industry according to the types of ecological industries, regional leading industries and development features. Specifically:

Preconditioned on circular usage of resources and ecological protection, the ecological agriculture mode relies on characteristic agricultural resources, farming culture and human settlements to vigorously develop green and organic agriculture, characteristic ecological planting and cultivation, biodiversity protection, planting of authentic medicinal materials and three-dimensional, circular ecological agriculture by promoting recycling, characterized, large-scale and high-end agriculture. It establishes and builds brands of characteristic ecological agricultural products, improves the market competitiveness of brands, boosts agricultural revitalization and benefits the public, thus realizing the unification of ecology with the benefits of agriculture, rural areas, and farmers. This mode is mainly applicable to areas with sound ecological environment where agriculture is the dominant function. For example, due to superior agricultural resources, Anji County (Zhejiang Province), Yuexi County (Anhui Province), Mengyin County (Shandong Province), Yuanyang Hani Terrace Heritage Area (Honghe Hani and Yi Autonomous Prefecture, Yunnan Province) and other regions have all provided vivid cases of advancing agricultural ecologicalization by energetically developing ecological circular agriculture, continuously improving agricultural industrialization level, and creating characteristic ecological agriculture brands.

With expansion, quality improvement, and transformation development at the core, the ecological industry mode enlarges and refines characteristic ecological industries through green transformation and the upgrading of traditional industries. It cultivates ecological industrial clusters and regional ecological brands with a base to improve ecological and environmental quality. This mode is driven by innovation and attracts a concentration of industries, capital, talent and technologies, fostering and developing resource-saving and environmental-friendly high-quality industries, pushing for a "rejuvenating" green and high-quality development. This mode is mainly suitable for areas where the ecological environment capacity is small and the development intensity is high, or where characteristic resources such as mountains, waters, forests, fields, lakes and grasses occupy a relatively large area. For example, Jiawang District (Xuzhou City, Jiangsu Province), Huaping County (Yunnan Province), Pingli County (Shaanxi Province), Huocheng County (Xinjiang Uygur Autonomous Region) and other regions have boosted the development of ecological industries by promoting industrial transition, optimizing the direction of industrial development, and successfully moving from sacrificing ecology to protecting ecology. The green development concept of "Lucid Waters and Lush Mountains are Invaluable Assets" has thus been applied effectively.

Relying on a good natural and ecological environment and unique human ecosystem, the ecological tourism mode holds sustainable development as its core concept, puts ecological environment protection at the premise, and follows the principle of harmonious development between humankind and nature. Following the idea of promoting tourism with protection, boosting development with tourism, and strengthening protection with development, this mode adopts eco-friendly methods to carry out ecological experiences, ecological education and ecological awareness activities. It innovatively builds new eco-tourism models and creates distinctive eco-tourism brands to fully leverage the driving force and catalytic integration of eco-tourism, thereby facilitating the coordinated development of regional economy and society. A win–win situation can thus be achieved for both tourism industry and ecological environment protection. This mode is mainly suitable for regions with a sound ecological environment and rich characteristic tourism resources. For example, by relying on solid characteristic natural and cultural resources, Yanqing District (Beijing), Danjiangkou City (Hubei Province), Xinxian County (Henan Province), Xunwu County (Jiangxi Province), Daocheng County (Sichuan Province) and Jiuzhaigou County (Sichuan Province) have all actively developed their ecological tourism industry to turn resource superiority into economic advantage. Practices and experiences that can be replicated and promoted have been formed in these areas.

Aiming to extend the industrial chain vertically and promote the integrated development of ecological formats, the "ecology+" compound industry mode integrates the concept of ecological development in industrial development, all-area-advancing tourism, ecological environment governance, scientific and technological finance, urban and rural development, rural revitalization, and related aspects. It promotes the coordinated development of large ecology and big data, modern agriculture, comprehensive tourism, and big health industry, cultivates new formats and new modes of integrated development of "ecology+" compound industry, constantly extends "ecology+" benefits, harnesses the synergy of industries, and realizes the diversified integration of new formats, the transformation and upgrading of industrial development, urban and rural environment, quality of life, as well as the improvement of overall urban status. This mode is mainly made for regions with excellent ecological resource endowment, great industrial foundation, and relatively concentrated innovation elements. For example, aiming to promote the integrated development of "ecology + industry" formats, regions such as Wudang District (Guiyang, Guizhou Province), Chongyi County (Jiangxi Province) and Fuliang County (Jiangxi Province) have boosted the coordinated development of industries, cultivated the new formats and new modes of the integration of primary, secondary and tertiary industries, and harnessed the synergy of industries, hence offering typical experience for the high-quality economic and social development, benefiting the people via ecology and developing ecology for the people.

3.3.3 Ecological Market Mode

Tasked with revitalizing ecological resources and promoting ecological market trading, this mode focuses on building market-based operations and market trading systems for ecological products and value realization. It seeks to establish ecological environment indicators and an equity trading mechanism based on the consumption of ecosystem services. By building market-oriented operations such as energy usage rights, carbon emission rights, pollution rights and water rights, it creates a market trading platform for ecological products with defined ownership and responsibilities. This will in turn balance interests between suppliers and consumers of ecological products, enable ecological resources to be asset-based, quantifiable, and tradable, and turn ecological assets into real money. This mode is mainly suitable for regions with rich ecological resources and developed systems. Relying on the market-oriented value realization mechanism of ecological products, Lishui (Zhejiang Province), Wanzhi District (Wuhu, Anhui Province) and Yongchun County

(Fujian Province), for instance, have developed and enriched ecological products and explored ways to establish a market-based trading platform. Through this, they have improved the market allocation of natural resources, providing sound experiences and references for effectively tackling difficulties in measurement, mortgage, trading and realization of ecological products.

3.3.4 Green Finance Mode

Facilitating the transformation of ecological benefits to economic benefits and advancing green finance are the main goals of this mode. By innovating in the areas of green financial products and financial measures, this ensures that ecological resources are equity-based, securitized, bond-based, and funded. It builds a financing guarantee system of ecological resources, turning green ecology into an 'endless pot of money' that generates wealth. This will allow for ecological environmental protection and sustainable economic and social development to grow hand in hand. This mode is mainly suitable for regions with a sound ecosystem and an advanced ecological economy. For example, Huzhou, Quzhou and Chun'an County (Zhejiang Province), have developed green finance in an innovative manner, improving the financing environment of green industries and green projects. They have also expanded financing channels, boosting ecological and environmental protection and restoration, providing hands-on experience for innovating financial products and serving the green industry.

Chapter 4
Typical Cases

Various regions across China, guided by Xi Jinping Thought on Ecological Civilization and the concept of "Lucid Waters and Lush Mountains are Invaluable Assets", have innovatively explored transformation paths and modes of "Lucid Waters and Lush Mountains are Invaluable Assets", as well as long-term mechanisms for value realization of ecological products. They have also proactively led public engagement in joint development of "Lucid Waters and Lush Mountains", delivered increased environmental benefits to local people, and have enabled people to benefit from the outcomes of "Invaluable Assets", thereby forming multiple hands-on experiences and vivid typical cases. In light of preliminary work foundation and on the basis of summarizing and analyzing the main measures of practice modes, the practice of "Lucid Waters and Lush Mountains are Invaluable Assets" has been further enriched and improved in detail and by layers in terms of background, specific practice, key achievements and inspirations. This forms the following ten typical cases for reference.

4.1 Luanchuan County (Henan Province)—Pushing Ahead with Mine Restoration to Promote Green Energy Transition

4.1.1 Background

Situated in the southwest of the city of Luoyang in Henan Province, and the hinterlands of Funiu Mountain in western Henan, Luanchuan County straddles two major water systems of the Yellow River and the Yangtze River, with 604 tributaries within its territory. It is characterized by intricately intertwined topography with many mountains and little farmland, as well as criss-crossing ravines and gullies. It is

H. Zhang et al., *Theory and Practice of "Lucid Waters and Lush Mountains Are Invaluable Assets" in China*, https://doi.org/10.1007/978-981-97-7340-4_4

thus known for its geomorphic features, embodied by the phrase: "Surrounded by four rivers, three mountains and two streams, ninety percent of the county is mountains, five percent is waters, and the remaining five percent is farmland". With a forest coverage rate of 82.4%, Luanchuan County boasts abundant natural resources, particularly spectacular mountains, beautiful waters and thick forests. Its mineral resources mainly include molybdenum, tungsten, iron, gold, silver, lead, zinc, and fluorite. Specifically, the county ranks first across China in terms of molybdenum metal reserves, is thus hailed as the "Molybdenum Capital of China", and ranks third nationwide in terms of tungsten metal reserves. Luanchuan County is a traditional industrial and mining county, which commenced development of mineral resources in the 1960s. Since 2008, Luanchuan County, guided by ecological conservation, has established a development strategy of "building an ecological county", conducting a series of tasks including mine restoration, ecological industry, ecological tourism, ecological agriculture, and ecological environment construction, and achieving remarkable results. On December 15, 2018, Luanchuan County was included in the second batch of "Lucid Waters and Lush Mountains are Invaluable Assets" Practice and Innovation Bases by the MEE of China.

4.1.2 Specific Practice

4.1.2.1 The Restoration and Governance of Ecological Environment Have Been Boosted

Luanchuan County has comprehensively advanced the restoration of mine environments, and made a strong push to restore ecological vegetation. The county has successively implemented a number of vegetation restoration projects at mining sites, slag discharge fields and tailings ponds. As of 2017, it had completed ecological governance of 767,700 m^2 of mines, and invested CNY100 million in carrying out ecological governance and restoration at 91 derelict tailings ponds across the county, with 3.555 million m^2 of ecological vegetation restored. The county has vigorously implemented the construction of ecological forestry by investing a total of CNY1.172 billion of financial funds in conducting major forestry ecological construction projects, such as afforestation by aerial seeding, returning farmland to forests, passageway verdurization, and village verdurization. The verdurization area is above 150 hm^2 each year, and the forest stock volume has reached 11 million m^3, ranking first in Henan Province. Furthermore, Luanchuan County has focused on developing ecological economic forests and establishing a forestry industry system. Through the implementation of land transfer, it has built 30,000 mu (or 2000 ha) of colored-leaf verdurization forests, and 67,000 mu (or 4466.67 ha) of passageway verdurization forests and economic mixed fruit forests, with such forests truly becoming the "money tree and fortune forest" for local people in the mountains. The county has made great efforts to implement comprehensive improvement of the rural environment, investing more than CNY3 billion to conduct a series of projects in this field, such as rural domestic wastewater control, garbage disposal and improvement of villages' appearance. 14 towns and one management committee across the county have all constructed domestic wastewater treatment plants, and the decontamination rate of urban domestic refuse reached 98% in 2017.

Promotion conference on all-area-advancing ecological restoration work in Luanchuan county
Photo Credit Natural Resources Bureau of Luanchuan County

4.1.2.2 Promoting the Green Development of Industrial Economy

Luanchuan County has implemented most rigorous systems for exploitation and protection of natural resources, adhering to the principle of "ecological development of industries", carrying out integration of resources and mining order readjustment at full stretch, and dividing mineral resource development zones and ecological tourism resource protection zones based on the distribution of these resources across the county, thus achieving economical and intensive ecological development of the mining industry. The county has conducted industrial green transformation and adopted multiple measures in this area, including special rectification of heavy metal-related sectors, special control of fluorine- and acid-related enterprises, and upgrading and control of gas-related enterprises. A total of 93 small mineral processing enterprises have been closed and dismantled, 292 poorly-managed and polluting small enterprises rectified and banned, and 38 enterprises audited for cleaner production. The county has strongly promoted cyanide-free mineral processing technology and in-depth treatment and reuse of industrial wastewater, and requested molybdenum and tungsten smelting enterprises to adopt carbon-free roasting technology and ultra-low emission control. Furthermore, it has optimized industrial structures, adjusted and upgraded any industrial layouts, scales and structures that do not conform to the functional orientation of the ecological environment, implemented a negative list system for industrial access, and has planned and constructed Luanchuan County's specialized park for tungsten and molybdenum new materials.

4.1.2.3 Deepening the Building of Ecological Agriculture Brands

Efforts have been made to build "Luanchuan Impression" as a regional agricultural product brand. 12 agricultural product processing enterprises have been identified to develop 81 products in six series—alpine miscellaneous grains, edible mushrooms, featured forest fruits, traditional Chinese medicinal materials, fruit wines and beverages, and traditional intangible cultural heritage. These have successively opened 18 flagship stores in Zhengzhou, Luoyang and other cities, built 11 online franchise stores, and set up on 6 auxiliary sales platforms, thus rolling out a diversified marketing system across the board. Furthermore, the county has raised awareness of the Junshan Pharmaceutical brand, utilized its advantages as the place of origin of dogwood, grifola, fructus forsythia and other traditional Chinese medicinal materials, and intensified the breeding of superior varieties of traditional Chinese medicinal materials and the planting of genuine medicinal materials, so as to enhance the influence of pharmaceutical brands. As a result, a number of sci-tech innovative firms like Bencaotang and Wokang Industrial Park have been formed. Furthermore, the county has built a kiwifruit industry chain. Underpinned by the planting of red kiwifruits and making use of the deep processing of kiwifruits, Luoyang Junshan Red Wine Industry Co., Ltd. has developed comprehensive agricultural parks to make kiwifruit wine series products. The kiwifruits and kiwifruit wines developed by the company have been certified as organic products, and its planting base has been identified by the agricultural department of Luoyang as "One Village, One Product" Characteristic Demonstration Park and Luoyang Fruit (Kiwifuit) Standard Park.

Agricultural products of "Luanchuan impression" *Photo Credit* Website of the People's Government of Luanchuan County

4.1.2.4 Elevating the Supporting Role of Ecological Tourism

Luanchuan County has fostered the tourism industry as the key industry of its county economy, and regarded "enriching the county by tourism" as one of the three strategies of its economic development. The county has initiated development of "Panoramic Luanchuan", putting forward an overall goal of constructing a new Luanchuan, a good place for travel and living brimming with vitality and kindness, and building a wellness city surrounded by the waters of Yihe River and the mountains of Luanchuan County. Relying on eight tourist attractions above 4A-level, including Laojun Mountain, Jiguan Cave, Chongdugou Ravine, and Baodu Village, it has driven surrounding villages to vigorously develop their tourism economy, facilitating the rural tourism economy according to the mode of "tourist attractions, folk towns and characteristic farms". Based on the construction of "county-wide verdurization" and "county-wide water systems", the county has simultaneously carried out 150 projects for building beautiful villages, and newly fostered and built a number of rural tourism demonstration villages, such as "Wellness Town" in Taowan Town, "Fragrant Lotus Town" in Qiupa Township, "Lingshang Family" in Boyunling Village, and "Danangou Manchu-style Leisure Resort Area" in Chengguan Town. Consequently, the pattern of "One Town, One Product" in 15 towns across the county has initially taken shape. On the basis of consolidating and improving the first batch of star villages for rural tourism such as Zhuangzi Village and Yangzigou Village, the county has implemented projects for creating A-level rural tourist attractions, and upgraded supporting facilities. For example, it has built multiple homestay brands, including a traditional homestay of Shimen Earth House, high-end homestays of

Manju·Shisanyue and Jingshui Shanju, and Xixi Miye, constructed 23 themed manors of photography and flower appreciation and pastoral nostalgia, and has developed 45 characteristic tourism specialized villages, thus establishing a new pattern of rural tourism with "leading scenic spots" and "leading villages" advancing together.

Chongdugou ecological tourism construction demonstration area *Photo Credit* Chongdugou Management Committee

4.1.2.5 Establishing and Improving the "Liangshan" Transformation Systems and Mechanisms

Luanchuan County has unveiled the Decision on Prioritizing the Implementation of the Strategy of "Building an Ecological County" to Accelerate Ecological Conservation and other relevant documents, giving top priority to the implementation of the strategy of "building an ecological county". It has set up the Ecological Conservation Commission of the County Party Committee, with the secretary of the County Party Committee serving as Director and relevant Party and government leaders serving as Deputy Directors. The Committee is responsible for the overall design, comprehensive coordination, integrated advancement, supervision and implementation of ecological conservation and environmental protection across the county. The Ecological Conservation Commission of Luanchuan County Party Committee has formulated the List of Responsibilities for Ecological Conservation and Environmental Protection Work of Member Units of the Ecological Conservation Commission of Luanchuan County Party Committee of the CPC, specifying the responsibility list of all units and the negative list for industrial development. Additionally, six working mechanisms have been established and improved, including promotion mechanisms for regular meetings on ecological and environmental protection work, multi-department coordination working mechanisms for illegal ecological and environmental cases, working mechanisms for supervision and assessment of ecological conservation, ecological compensation mechanisms, working mechanisms for special audits of environmental protection work, and accountability mechanisms for ecological and environmental damage. The establishment and improvement of "Liangshan" transformation systems and mechanisms in Luanchuan County have expedited ecological environment protection and the implementation of major projects, achieved "industry ecologicalization and ecology industrialization", and constantly guaranteed ecological environment improvement, ecological economic development and ecological benefits to local people such as increases in income, thereby promoting the continuous transformation of lucid waters and lush mountains into invaluable assets.

Ecological planting industry in Taowan town of Luanchuan county *Photo Credit* Luanchuan County Bureau of Agriculture and Rural Affairs

4.1.3 Key Achievements

The industrial and mining sector has long been the only key industry in Luanchuan County, causing severe damage to the ecological environment. In recent years, Luanchuan County has resolutely promoted the green transition, carrying out green transformation of the industrial and mining sector with tenacious determination. The whole county has been mobilized to restore the ecological environment, develop all-area-advancing tourism, build a regional brand of "Luanchuan Impression", and deepen the construction of ecological agriculture. As a result, the county has realized a transformation from sacrificing ecology to protecting ecology and from "living off of mountains through mining" to "living off of mountains through tourism", thereby practicing the concept of "Lucid Waters and Lush Mountains are Invaluable Assets" in a vivid way.

4.1.3.1 Continuously Optimizing the Quality of the Ecological Environment

Through the restoration of mine environments, Luanchuan County has effectively improved the ecological environment of mountains. Attaching great importance to environmental protection work, CMOC Group Limited has been rated as the "National Advanced Production and Construction Unit of Water and Soil Conservation Work" and the "Ecological Well-off Demonstration Base of China". Industrial green transformation projects have been consistently implemented. Over the past three years, the county as a whole has reduced chemical oxygen demand by 849.68 tons, ammonia nitrogen by 75.1 tons, sulfur dioxide by 196.24 tons (excluding smelting enterprises), and nitrogen oxides by 94.28 tons, achieving strong results in main pollutant emission control. The number of days with good environmental and air quality remained over 320, and the eco-environmental quality index (EEQI) remained above 80, increasing year by year.

4.1.3.2 Changing the Appearance of Villages from the "Old Look" to the "New Look"

The improvement and development of villages' appearance has enabled villages to embark on a path to sustainable development featuring utilization and protection, putting equal emphasis on development and protection. From adobe houses in the 1940s and 1950s before rural tourism was developed to today's clean village appearance, centralized treatment of wastewater and establishment of garbage collection and disposal system, the "old look" has been changed into the "new look".

4.1.3.3 The Development of the Ecological Tourism Industry Has Been Full of Vitality

Luanchuan County has realized the exploitation of urban and rural scenic areas, countywide development of scenic areas, and quality improvement as a good travel destination. In 2018, the added value of the tourism industry in Luanchuan County accounted for 16.3% of its GDP, with the whole county receiving 14.876 million tourists, and the total tourism revenue reaching CNY8.74 billion, directly lifting more than 12,000 people in financial difficulties out of poverty, and increasing the per capita annual income by CNY15,000. The ecological tourism industry has truly become a strong support for the county economy, and an industry that drives local people to eliminate poverty and achieve prosperity.

Beautiful Luanchuan county *Photo Credit* Website of the People's Government of Luanchuan County

4.1.3.4 Increasing Prominence for the Effects of Featured Brands

Driven by the "Luanchuan Impression" brand, the county has built a total of 23 production and processing lines, developed 27 poverty alleviation bases, and led 1751 poverty-stricken households and 5250 natives to increase their income and achieve prosperity. Through land transfer, labor employment, free technical training, and product recycling, Junshan Red Wine Industry Co., Ltd. has led more than 220 farmer households and enabled famers to transform from "relying on mines for survival" to "relying on agriculture for income increase", embarking on the sustainable development road set out by "Lucid Waters and Lush Mountains are Invaluable Assets".

4.1.4 *Inspirations*

Luanchuan County has implemented the strategy of building an ecological county, solidly pushed forward relevant work to construct "Liangshan" bases, and strived to transform the requirements of construction of socialism with Chinese characteristics in the new era into the endogenous motivation and concrete practice of construction of "Lucid Waters and Lush Mountains are Invaluable Assets" bases. It has optimized industrial structures, conducted industrial green transformation, and continued to promote economic and social transformation and development, thus facilitating the integrated development of agriculture and tourism with the creation of an all-area-advancing tourism demonstration county as the lever. The concept of countywide development is deeply rooted among the public, and remarkable achievements have been made in benefiting local people and enriching farmers. As a result, the transformation mode of "Lucid Waters and Lush Mountains are Invaluable Assets" with the characteristics of Luanchuan has been gradually formed, and the typical experience and successful mode are worth promoting nationwide.

4.2 Youyu County (Shanxi Province)—Creating a New Paradigm to Control Desertification and Enrich People

4.2.1 *Background*

Located at the border between the two provinces of Shanxi and Inner Mongolia, Youyu County is considered as the north gate of Shanxi Province. It has a temperate continental climate with a large diurnal temperature difference and many windy days. Its soil texture is relatively rough and it is prone to wind erosion, water erosion, desertification and other phenomena. In the early days after the People's Republic of China was founded, there were only 8000-mu (or 533.33 ha) remnant forests in the whole county, with the forest coverage rate being less than 0.3%, and the land desertification area accounting for 76%. Local people suffered from frequent sandstorms. In 1949, in the face of barren mountains and rivers where sandstorms frequently occur, the first secretary of the county Party committee led the whole county to start a magnificent feat to "plant trees and reconstruct mountains and rivers". Since then, successive secretaries have worked towards one goal with one blueprint, adhering to the achievements view of "I do not need to be the one to succeed" to unite and guide the cadres and the people of the whole county. Over nearly 70 years of relentless efforts in afforestation, desertification control and ecological improvement, the county has vigorously practiced the concept of "Lucid Waters and Lush Mountains are Invaluable Assets". It has controlled 2 million mu (or 133,333.33

ha) of desertificated land, with the forest verdurization rate in the whole county reaching 56%. The former "barren land" has become the well-known "Oasis beyond the Frontier", nurturing and forming the precious Youyu Spirit of "serving the people wholeheartedly, working hard to rise to challenges, and making unremitting efforts for long-term benefit".

President Xi Jinping has given important instructions on the Youyu Spirit on five occasions, noting that "The Youyu Spirit is a valuable asset, which must be vigorously studied and promoted." Youyu County has won numerous honorary titles, including outstanding contributor for the construction of Three-North Shelterbelt Project, national advanced verdurization collective, national ecological civilized county for water and soil conservation, national ecological conservation demonstration county, "Lucid Waters and Lush Mountains are Invaluable Assets" Practice and Innovation Base, and outstanding contributor for 20th anniversary of Iseeforestry.

Ecological scenery in Youyu county *Photo Credit* Website of the People's Government of Youyu County

4.2.2 Specific Practice

4.2.2.1 Conducting Afforestation to Consolidate the Foundation of the Ecological Environment

Youyu County has combined engineering measures and biological measures, adopting the principle of planting trees, shrubs and grasses on suitable land to effectively control water and soil erosion within its watersheds, and to improve forest and grass coverage rate. Given the reality of late soil thawing in spring and the short period suitable for tree planting, the county has introduced an approach of tree planting in spring and autumn and soil preparation in autumn according to its local conditions. The county has mobilized professional teams, private organizations, enterprises, cadres and local people to voluntarily plant trees, so as to effectively stimulate the vitality of ecological development. All key afforestation areas across the county have been enclosed with barrier nets to strengthen the protection of ecological forests, natural reserves, water conservation areas, important geological relics and biological species resources, thus safeguarding biodiversity and the original ecology of plants.

Verdant Canghe river *Photo Credit* Website of the People's Government of Youyu County

4.2.2.2 Innovating Systems and Mechanisms to Reinforce a Long-Term Guarantee for the "Liangshan" Transformation

Youyu County has comprehensively deepened the reform of the collective forestry rights system, explored ways to establish a green development fund, and motivated the enthusiasm of main market players to invest in the forestry construction, in a bid to promote the large-scale and intensive development of the forestry industry. The county has strictly implemented the *Regulations on Responsibilities for Environmental Protection Work in Youyu County (For Trial Implementation)*, and the Party and government leaders have personally presided over and assumed overall responsibility for environmental protection work, thus monitoring, supervising and evaluating environmental protection targets in a stringent manner. The county has conducted strict environmental protection access to projects, and based on the development of the green economy, held the threshold of project access with a focus on park-oriented industrial layouts, agglomerative development and clean production of enterprises. It has clarified the "six red lines" for environmental protection approval of construction projects. The county has cracked down on environmental violations, paying special attention to strengthening the coordination between administrative law enforcement departments and judicial departments for environmental protection, thereby forming a strong work synergy in law enforcement supervision of environmental protection. It has reinforced the protection of ecological resources, set up forest police stations, forest fire brigades and forest resources protection patrol teams, and built national standard stations and central monitoring and reporting sites.

4.2.2.3 Developing Green Ecological Industries to Enhance the "Liangshan" Practice Connotation

Youyu County has vigorously developed ecological planting by promoting the characteristic planting of high-quality oat, oilseed rape, flax and quinoa, pushing forward the standardized planting of traditional Chinese medicinal materials, and building three major traditional Chinese medicinal materials planting bases of superior radix isatidis, radix sophorae flavescentis and scutellaria baicalensis. The county has implemented a subsidy and reward policy for grassland ecological conservation across the board, and set up an animal husbandry industrialization system integrating forage grass planting, large-scale breeding, slaughtering and processing. Furthermore, by taking the sea buckthorn industry as the priority for the development of economic forests, the county has transformed low-yield sea buckthorn forests, introduced high-quality big-fruit sea buckthorns, planted Chinese small-fruit sea buckthorns, and has brought in large-scale sea buckthorn enterprises. The county has also energetically developed the nursery stock industry, listing nursery gardens, green belts, ecological parks, scenery lines and demonstration areas as the focus of ecological conservation. Finally, the county has proactively created an advantageous agricultural product processing system, expanded the industry chain of agricultural products making use of large-scale ecological agriculture, and vigorously developed the green food processing industry.

4.2.2.4 Advancing Regional Construction to Build the "Liangshan" Ecological Brand

Youyu County has taken its ecological culture and tourism development zone as a core area of the national all-area-advancing tourism demonstration zone, highlighting the ample atmosphere of cultural and creative arts in Youyu County through an ingenious layout. The county has developed at full stretch the production of commercial paintings and the creation of famous paintings, thus striving to form a base for oil painting production, photography creation and printmaking with Youwei Ancient City at the center. Renowned domestic experts have been engaged to offer advice and suggestions on the green development and the construction of development zones in Youyu County. The county has collaborated with Anji County in

Zhejiang Province, Yan'an City in Shaanxi Province, Saihanba Forest Farm in Hebei Province, and Aksu City in Xinjiang Uygur Autonomous Region to establish an "Alliance for Ecological Culture and Tourism Development", and to jointly release the *Youyu Green Declaration*, significantly enhancing its visibility and reputation.

Autumn harvest *Photo Credit* Website of the People's Government of Youyu County

4.2.2.5 Balancing Urban and Rural Development to Improve the Quality of the Human Settlement Environment

A large number of municipal facilities have been built and put into operation across the county. These include co-generation, natural gas access, pipeline networks of wastewater treatment, garbage treatment and sewage discharge, and a transportation network with expressways and national and provincial highways as its main trunks, and county and township roads as its branch lines. Youyu County has implemented a rural human settlement improvement project and a rural drinking water consolidation and improvement project, gradually carried out treatment of rural domestic garbage and wastewater, and vigorously promoted pilot work on renovating, activating and utilizing idle and dilapidated house sites, significantly improving the human settlement environment.

The most beautiful county in China—Youyu county *Photo Credit* Website of the People's Government of Youyu County

4.2.2.6 Fostering an Ecological Concept to Advocate Ecological Ethics Culture

Youyu County has performed its primary responsibility in the ideological sphere, effectively integrated the Xi Jinping Thought on Ecological Civilization into all areas of daily life and work, developed environmental protection groups, set up green volunteer teams, encouraged the creation of green literature, calligraphy and paintings and photography works, promoted voluntary afforestation activities, established a public hearings system for environmental protection issues, and has carried out other public engagement activities. The county has strengthened policy guidance, and has also offered supporting services and other preferential services to those actively developing the green, low-carbon and cyclic economy, so as to form a science-based and effective mechanism with the guidance, drive and constraint of ecological culture.

4.2.3 Key Achievements

4.2.3.1 Achievements in Afforestation

Over the past seven decades, the county as a whole has controlled 2 million (or 133,333.33 ha) of desertificated land, with the forest verdurization rate reaching 56%. The forested area of the whole county currently accounts for 1.7 million mu (or 113,333.33 ha), sand and dust storm weather has been 80% less than that in the early days after the People's Republic of China was founded, overland runoff and river sediment concentration have decreased by 60% from that before afforestation, and the annual average water evaporation quantity in the farmland forest network has reduced by 8.8% compared with that of the wilderness.

4.2.3.2 Achievements in the Improvement of the Ecological Environment

In 2018, Youyu County had 292 days with good air quality, representing an increase of 9 days year-on-year. Meanwhile, it recorded an average concentration of sulfur dioxide of 24 $\mu g/m^3$, an average concentration of PM of 33 $\mu g/m^3$, and a comprehensive index of ambient air quality of 4.51, down by 45.5%, 8.3% and 13.9% on a year-on-year basis respectively, ranking first in Shuozhou City.

4.2.3.3 Achievements in the Development of Green Industries

"Youyu oat rice" has applied for a national geographical indication. Currently, Youyu County has transformed 40,000 mu (or 2666.67 ha) of low-yield sea buckthorn forests, introduced 4300 mu (or 286.67 ha) of high-quality big-fruit sea buckthorns, planted 24,300 mu (or 1620 ha) of Chinese small-fruit sea buckthorns, brought in 12 large-scale sea buckthorn enterprises, and created more than 30,000 tons of various products annually. It has cultivated over 450 large-scale nursery gardens with an annual output value of seedlings at CNY238 million, and has built the largest production base of pinus sylvestris nursery stock in North China.

4.2.3.4 Achievements in the Development of Ecological Tourism

Youyu County is the first 4A-level ecological scenic spot named after a county in China, part of the first group of national all-area-advancing tourism demonstration zones, and the first provincial-level development zone dedicated to the development of the ecological and cultural tourism industry in Shanxi Province. In 2019, Youyu County received 3.883 million tourists, realizing a tourism revenue of CNY3.56 billion, representing year-on-year increases of 33.8% and 31.9% respectively.

4.2.3.5 Achievements in the Development of the Ecological Concept

In 2019, through the campaign of "setting examples, promoting demonstration practices and establishing mechanisms", Youyu County has created a total of more than 230 ecological civilized bodies, enterprises and public institutions, communities and villages with a creation rate of 80%, shaping an excellent style described as "I'm exposed to the ecology, and the ecology is kept in my mind". As a result, green travel accounted for 90% of that taken by the public, the availability of energy-saving and water-saving apparatus reached 81%, and the proportion of green government purchase was 91%.

Shahukou ecological tourism *Photo Credit* Website of the People's Government of Youyu County

4.2.4 Inspirations

Youyu County has steadfastly built up confidence in the path, theory, system, and culture of the socialist approach to ecological civilization, adhered to the concept of "Lucid Waters and Lush Mountains are Invaluable Assets", followed green development modes and lifestyles, and strove to transform the requirements of construction of socialism with Chinese characteristics in the new era into the endogenous motivation and concrete action of construction of "Liangshan" practice and innovation bases. This has gradually formed a transformation mode of "Lucid Waters and Lush Mountains are Invaluable Assets" with its own characteristics—"Youyu Mode", in terms of ecological environment protection and green development. From the formation of the Youyu Spirit, it can be found that, despite an ever-changing development environment, work priorities and work methods, every term of the county Party committee has firmly believed one truth over the past seven decades: planting trees is the fundamental solution to meet the needs of the masses for survival and development, and planting trees means political concerns and the common aspiration of the people, which is both a responsibility and a mission.

4.2.4.1 Strengthening Ecological Protection and Restoration to Preserve and Increase the Value of Natural Assets

Based on its local conditions, Youyu County has put forward different afforestation ideas according to development requirements at different periods. In the 1950s, the county adhered to the practice that "trees shall be planted in windy areas to make them green first", thus beginning a path to afforestation. In the 1960s, in light of the requirement that "trees shall be planted in windy areas to control windblown sand", the county highlighted the control of windblown sand. In the 1970s, according to the concept that "trees shall be planted on vacant land, and holes shall be filled up then", the county made strong efforts to build the shelter forest system. In the 1980s, following the principle that "appropriate trees shall be planted on suitable land in a reasonable manner, and three species of pine trees shall be introduced", it further improved the quality of afforestation. In the 1990s, the county introduced a three-dimensional afforestation mode by adhering to the approach that "trees and shrubs shall be planted in a mixed and reasonable manner to build green barriers". Entering the twenty-first century, the county proposed the development concept that "trees shall be planted contiguously after returning farmland to forests to make mountains and rivers everywhere more beautiful". For now, the county has insisted on the principle that "lucid waters and lush mountains beyond the frontier shall enrich the locals". In the long-term practice of green development, Youyu people have explored a scientific and effective ecological governance mode and approach to afforestation, which have not only significantly improved the local ecological environment, but also provided precious referential experience for the ecological restoration and sustainable development of alpine and drought regions.

4.2.4.2 Fully Exploring Characteristic Industries to Push Forward Green Development

Based on the realities of its situation, Youyu County has accurately analyzed the realistic requirements of the scientific connotation of the concept of "Lucid Waters and Lush Mountains are Invaluable Assets" for regional economic and social development, highlighting the value embodiment of an "ecological Youyu" in terms of the development of under-forest economy and the horse industry. Utilizing abundant forest resources, the county has developed forestry productivity, taken the sea buckthorn industry as a priority for the development of economic forests, vigorously developed the nursery stock industry, and facilitated the processing of minor grain crops, tartary buckwheats, sea buckthorns, dehydrated vegetables, flaxes and other advantageous agricultural products. In addition, it has cultivated a horse industry base and a horse breeding center, launching an international horse racing open tournament, to improve the influence of Youyu County both at home and abroad. At present, Youyu County has built the largest horse industry base in China, integrating a breeding center, racing center, audition center and training center. These endeavors have strongly promoted the ecological and cultural tourism of Youyu County and enabled the ecological culture and the horse industry to promote and depend on each other, thereby forming a development pattern featuring interdependence and mutual promotion.

4.2.4.3 Energetically Carrying the "Youyu Spirit" Forward by Relying on Museums, Exhibition Halls and Other Carriers

Youyu County has implemented the core value of Chinese socialism, and proactively advocated the concept of ecological civilization. With the support of the Youyu Museum, the Exhibition Hall of Youyu Spirit, *Youyu and Her Secretaries of the County Party Committee* and other carriers and news media, the county has actively publicized and carried forward the Youyu Ecological Culture Thought and the Youyu Spirit, such as the concept of "working hard from generation to generation, and unswervingly adhering to a blueprint", as well as the "achievements view that all communists shall make unremitting efforts for long-term benefit". The county has guided many cadres and local people to spread ecological awareness across the county, so as to effectively integrate Xi Jinping Thought on Ecological Civilization into all areas and aspects of daily work and life, and to push for the whole society to form a good social morality with everyone respecting nature, following its ways, and protecting it.

4.3 Xunwu County (Jiangxi Province)—Leveraging Ecological Advantages to Comprehensively Develop Rural Tourism

4.3.1 Background

As the headstream of the Ganjiang River, Dongjiang River and Hanjiang River, Xunwu County (Ganzhou City, Jiangxi Province) represents an important component of ecological barriers in South China and a key national ecological functional zone, where the late Chairman Mao Zedong conducted the "Investigation of Xunwu" in 1930. During the second revolutionary civil war (1924–1927), Mao Zedong, Zhu De, Deng Xiaoping and other proletarian revolutionaries of the past generation engaged in great revolutionary practice here. In particular, the late Chairman Mao conducted an investigation for nearly a month in Xunwu County in May 1930, wrote two magnificent works, i.e. *Report From Xunwu* and *Combat Book Worship*, and proposed such celebrated theses as "no investigation, no right to speak". Therefore, Xunwu County has become one of the birthplaces of the CPC's ideological line of seeking truth from facts. Meanwhile, as the source of the Dongjiang River, Xunwu County enjoys strong ecological advantages, with the area of Dongjiang River source basin within its territory accounting for 56% of the area of Dongjiang River basin in Jiangxi

Province. In recent years, Xunwu County has implemented the concept of "Lucid Waters and Lush Mountains are Invaluable Assets", fully leveraged its ecological advantages as the source of Dongjiang River on the premise of green rising, and made good use of the "Investigation of Xunwu" and other gold-lettered signboards of red revolution in a flexible manner. The county has followed "five principles", realized "four breakthroughs", and pushed forward the rapid development of "red + green" characteristic tourism, thus achieving both ecological and economic benefits.

Memorial hall for investigation of Xunwu *Photo Credit* Website of the People's Government of Xunwu County

4.3.2 Specific Practice

4.3.2.1 Adhering to Planning Guidance to Coordinate Development Patterns

On the construction of scenic areas, Xunwu County has followed the principle of "planning before construction". For all projects, planning and design schemes must be formulated, and the requirement of "protection in development and development in protection" must be implemented. All projects shall be submitted to the county planning committee for discussion and approval, and no project shall be launched without approval. The county has taken a broad view of all areas across the county and analyzed the overall situation. It has formulated the *Master Plan on the All-area-advancing Tourism in Xunwu County*, providing a basis for the overall development positioning and layout of rural tourism. The county has also constructed every scenic spot according to plans. Development enterprises have been required to conduct development and construction in strict accordance with the master plan and detailed plan of the scenic area, formulate the master plan and regulatory detailed plan for all tourism projects under construction or already in development and operation, and put in place ecological protection measures to ensure the quality of project development and construction. In the meantime, tourism enterprises shall adjust project contents based on their development needs, which must be submitted for approval before implementation. For example, the Qinglong Rock Tourist Resort intends to add an experiential new format project in Danxia Scenic Area in order to increase its attractiveness and retain customers. It must obtain the consent of relevant departments of the county before implementation, in a bid to ensure standardized and orderly construction, protection and development of the scenic area.

4.3.2.2 Giving Priority to Protection to Consolidate the County's Ecological Foundation

Xunwu County has regulated investors' development behaviors. When signing relevant development agreements, the county has specified that investors must strictly implement the *Environmental Protection Law of the People's Republic of China*, the *Water Pollution Prevention and Control Law of the People's Republic of China*, the *Regulations on the Administration of Construction Project Environmental Protection*, the *Regulation on the Implementation of the Water and Soil Conversation Law of the People's Republic of China*, among other laws and regulations. The county has raised the requirement that tourism projects shall be developed and constructed according to plans, conducting real-time supervision on whether the development of scenic areas has passed the environmental impact assessment (EIA) and on whether there is damage to tourism resources and ecological environment, and should give top priority to the protection of ecological resources, so as to avoid repeated construction and resource-destructive construction. The county has also pursued win–win development. On the basis of strictly holding the upper limit of resource consumption, the bottom line of environmental quality and the red line of ecological protection, it has fully leveraged its resources and built characteristic tourism to ensure the reasonable, efficient development and utilization of resources. For example, Huakaile Ecological Tourism Scenic Area, formerly an abandoned fruit mountain, has been built into a tourism destination mainly featuring ecological leisure, water conservation demonstrations, science popularization education and health resorts through construction and creation by enterprises, thus enabling the transformation of "abandoned fruit mountain—nursery garden—scenic area", and achieving the goal of creating a win–win situation for ecology and the environment.

4.3.2.3 Maintaining the Dominant Role of Projects to Expand Tourism Routes

Xunwu County has specified dispatched county leaders and responsible units of projects, mapping out annual construction plans and goals to ensure the construction progress of projects. The county has also intensified work dispatching and taken various measures to ensure the smooth construction of projects, such as taking the initiative to coordinate with relevant departments to solve problems encountered during construction, and conducting field investigation on a regular or ad-hoc basis. Furthermore, the county has introduced incentive policies to strengthen guarantees. It has unveiled the *Incentives on Promoting the Development of All-area-advancing Tourism in Xunwu County (For Trial Implementation)*, in a bid to offer rewards and support in terms of land security, tax support, brand creation, key formats, infrastructure, and market development, and to actively ensure the successful development of the tourism industry. The county has also proactively created featured tourism products. It has fully leveraged local unique tourism resources to develop influential characteristic formats. For example, the Qinglong Rock Tourist Resort has made full use of the advantages of big stones in Linshibei Village to build Lingshi Hot Spring Project, which has attracted numerous tourists. Meanwhile, it has built an aquatic boutique route, and taken full advantage of the high-quality water resources in Xunwu County to develop such characteristic projects as Danxia sightseeing and hot spring wellness at Qinglong Rock Tourist Resort, canyoning at Shikongzhai Scenic Area, and water recreation at Douyan Lake, which have been popular among tourists.

Shikongzhai scenic area *Photo Credit* Website of the People's Government of Xunwu County

4.3.2.4 Utilizing Resources to Cultivate Tourism Brands

Xunwu County has reinforced service guidance, conducted regular and ad-hoc field investigation, actively recommended start-ups to higher authorities, and put in place reward and supplement measures for brands, enhancing the enthusiasm of tourism enterprises to improve the quality of scenic areas, and expediting the creation of tourism brands in Xunwu County. Currently Xunwu County has successfully created one national 4A-level scenic area (Qinglong Rock Tourist Resort), one 4A-level rural tourist spot (Huakaile Ecological Scenic Area), three 3A-level rural tourist spots (Shikongzhai Tourism Scenic Area, Xiangshanzeng Resort Chateau and Douyan Lake Tourism Scenic Area), and two flagship stores of Hakka cuisine (Xunwu Hotel and Xunwu Yuxianglou Hotel).

Qinglong rock scenic area *Photo Credit* Website of the People's Government of Xunwu County

4.3.2.5 Constantly Improving Elements to Highlight Characteristic Advantages

Xunwu County has actively explored special cuisines locally, and developed Xunwu feasts including feast of the Hakkas, Tofu feast, and whole lamb feast. Among them, "Xunwu Feast of the Hakkas" at Xunwu Hotel has been rated as a famous feast in Jiangxi. The Xunwu tourist distributing center has almost been completed, and relevant equipment is being prepared for installation. The Qinglong Rock Hot Spring Hotel, Dongjiangyuan Hot Spring Hotel, Vienna International Hotel and Guangxun International Hotel built under the standard of four-star hotels have also all completed the construction of main structures. The county has pushed ahead with the construction of homestays. In view of diverse resources, it has built homestays in different styles, such as Qinglong Rock Hot Spring Homestay, Huakaile Garden Homestay, and Jinguigu Forest Homestay. Huakaile Garden Homestay is constructed in a scattered way on different hills and lakesides, maintaining the style of integration with the scenery in the scenic area to the greatest extent possible. During the activity of "Give Ganzhou a Thumb up" initiated in 2020, it won the first place as the "most desirable homestay hotel". Furthermore, the county has also promoted the construction of tourist toilets. While advancing the benchmarking construction of A-level tourist toilets, it has conceived and designed distinctive tourist toilets based on the unified style and terrain of each scenic area.

Xiangshan tourist resort *Photo Credit* Website of the People's Government of Xunwu County

4.3.3 Key Achievements

4.3.3.1 Achieving a Breakthrough in the Number of Tourists

In 2019, the number of tourists received and tourism revenue in the whole county exceeded 2 million and CNY1 billion respectively. Specifically, the county registered 2.0677 million and CNY1.282 billion in number of tourists received and the tourism revenue, representing an increase of 94% and 189.7% year-on-year respectively, effectively driving rural economic development.

4.3.3.2 Achieving a Breakthrough in Brand Creation

In June 2020, the Qinglong Rock Tourist Resort was successfully built into a 4A-level scenic area, solving the issue of the lack of 4A-level scenic area in Xunwu County, representing a historic breakthrough. It has also further increased the popularity of Xunwu County, and built a strong brand at the source of the Dongjiang River in Xunwu County.

4.3.3.3 Achieving a Breakthrough in Building a Boutique Route

In the last five years, Xunwu County has introduced a total of ten tourism projects, with the amount of tourism investment exceeding CNY10 billion in cumulative terms. It has built an aquatic boutique tourism route from the Qinglong Rock Tourist Resort to the Shikongzhai Tourism Scenic Area, Douyan Lake Tourism Scenic Area and Huakaile Ecological Scenic Area. Meanwhile, thanks to the investment from enterprises, it has further protected and improved the ecological environment of scenic areas and surrounding areas.

4.3.3.4 Achieving a Breakthrough in Integrated Development

In recent years, Xunwu County has made full use of its ecological advantages, and actively promoted the development of tourism with agriculture and tourism with wellness, among other industries. It has constructed three wellness and health projects, built about 50 characteristic agricultural picking gardens, and developed an aroma series of products of cinnamomum camphora, dried passion fruits and other featured tourist commodities, pushing for farmers to increase their income and achieve prosperity.

Dongjiangyuan ecological town in Xunwu county *Photo Credit* Website of the People's Government of Xunwu County

4.3.4 Inspirations

Leveraging its characteristic advantages, Xunwu County has planned all-area-advancing tourism, and developed tourism products according to development goals and positioning, so as to shape the image of rural tourism of "arousing nostalgia while enjoying landscape sceneries", and to build tourism brands with highlights and characteristics. Xunwu County has implemented the concept of "Lucid Waters and Lush Mountains are Invaluable Assets" in every aspect of the tourism industry. Relying on its superior natural ecological resources and high-quality agricultural industries, Xunwu County has proactively pushed forward the integration of tourism with agriculture and culture, regarded the environment as an important component of the tourism product system, and highly valued the ecological factors contained in tourism products. The county has comprehensively protected the local ecological environment, taken full advantage of abundant forest resources, high-quality water resources and a good ecological environment, developed ecological tourism products with local features, and overcome the short-sighted behavior of ignoring environmental protection and ecological environment in the development of tourism, thus achieving the sustained and healthy development of rural tourism. Focusing on the goal of making tourists happy, comfortable and satisfied, it has effectively accomplished six elements of tourism, ensured the reasonable allocation of facilities and equipment, and supplied distinctive catering, comfortable accommodation and secure tourism products, thereby transforming ecological benefits into economic benefits, and promoting the development of rural tourism.

4.4 Xinxian County (Henan Province)—Building Ecological Industries to Continuously Unleash Green Momentum

4.4.1 Background

Located in the hinterland of Dabie Mountain in South Henan, and straddling the Yangtze River and Huaihe River, Xinxian County is the source of the Huanghe River, Bailu River, Zhaihe River, Zhugan River, Daoshui River and Jushui River, and also the most important ecological barrier in the southeast region of Henan Province and a key national ecological functional zone, hence enjoying crucial ecological status. Due to the impact of war and the constraints of natural conditions, Xinxian County once lagged behind in development. It was one of the key counties included in national plans for poverty alleviation through development, and one of the key counties for poverty reduction in Dabie Mountain contiguous areas of extreme poverty. In recent years, Xinxian County has vigorously implemented a strategy of building itself into an ecological county, firmly established a philosophy of "red guidance and green development" with a focus on the development orientation of "scenic red city and healthy Xinxian", and advocated ecological awareness of "regarding mountains as the father, waters as the mother, and forests as the son". The county has remained committed to the "construction of mountains and forests", played an "ecological card", and taken on a "characteristic road". It has continuously advanced the transformation of "ecological capital" to "capital enriching people", and worked to open the transformation channel of "Lucid Waters and Lush Mountains are Invaluable Assets", thus achieving remarkable results. As the rural revitalization demonstration county in Henan Province, Xinxian County has won dozens of national-level honors, including national sanitary county for five consecutive years, national garden county for four consecutive years, national civilized city for three consecutive years, and national ecological conservation demonstration county.

Scenic Xinxian county *Photo Credit* Website of the People's Government of Xinxian County

4.4.2 Specific Practice

4.4.2.1 Maintaining an Excellent Ecological Environment All year Around by Strengthening Protection and Governance of the Ecological Environment

Xinxian County has worked hard to implement projects for "blue sky, clear water and clean villages", vigorously promoted links among the county, towns and villages, and completed the "three-level integration" of gas, water and soil. As a result, the environmental quality of the county has continued to improve steadily. The control of rural nonpoint pollution has been advanced across the board, and a supervision, control and restoration system for soil pollution has been gradually established, with zero emissions of fertilizers and pesticides. The protection and construction of "one area and six parks" including Dabie Mountain (Xinxian County) National Park, Xiangshan Lake National Water Conservancy Scenic Area, and Xiangshan Lake National Wetland Park has been carried out at a steady pace. The county has implemented a permanent commercial ban on felling, logging and retrofitting of natural forests and public welfare forests, registered and archived mountains and rivers, and protected 1500 ancient and famous trees by labeling them. 991,000 mu (or 66,066.67 ha) of public welfare forests have all been covered by forest insurance, and there have been no major forest fires for 40 consecutive years. The county has scientifically delineated the red line of ecological conservation, identified "three areas and three lines", and strictly defined the scope of access. It has delineated five water function areas and 18 protection areas of centralized drinking water sources, identified banned areas, restricted areas and suitable areas for livestock and poultry breeding, with the protected areas accounting for 46.3% of its land area.

Yufan village by Xiangshan lake *Photo Credit* Website of the People's Government of Xinxian County

4.4.2.2 Ecological Industries Boost Comprehensive Development Countywide Through the Practical Explorations of "Liangshan" Transformation

Xinxian County has proactively developed the green processing industry of agricultural products, cultivated 30 provincial- and municipal-level leading enterprises of agricultural industrialization, formulated 19 local standards for featured agricultural products, certified 13 agricultural products as "pollution-free, green, and organic agricultural products" and labeled them with "geographical indication", covering an area of 257,000 mu (or 17,133.33 ha). The green industry, dominated by emerging sectors, has developed rapidly. In the first half of 2012, Henan Junton Vehicle Co., Ltd. recorded USD44.537 million in the export value of special-purpose vehicles, and expanded its export markets from Russia, Central Asia and other countries and regions to Europe, South America and other middle- and high-end markets. Many of the county's projects have been progressing with success, including the demonstration project of ten thousand cows, comprehensive processing of natural drinks favored by young people, and functional fruit and vegetable juices. The county has promoted the transition and upgrading of the coal, salt and chemical industries as well as other traditional industries, striving to build the largest fine chemical base in Central China.

With a strong development momentum of its ecological tourism industry, a development pattern of rural tourism involving "9 towns and 18 bays" from one point to the entire region has been built. The county has pushed forward joint construction and sharing, and constantly expanded entrepreneurship and employment paths. It has set up a special fund of CNY10 million for improving the rural living environment, and provided 1723 public welfare posts such as cleaners and ecological environment conservators, and over 6000 temporary posts, effectively driving employment and increases in income. The county has adhered to integrated development, and continuously magnified the effectiveness of poverty alleviation by developing industries. It has built 86 entrepreneurship demonstration bases covering areas such as agricultural sightseeing, fruit and vegetable picking and culinary experiences, and organized various cultural and tourism activities, festival activities, and literature and artistic creation activities themed around Dabie Mountain. It has raised the popularity of the "season of enjoying flowers" tourism brand, and cultivated more than 70 rural tourism operation units, tourist homestays, leisure and sightseeing parks, and characteristic rural tourism villages.

Xihewan ecological tourism scenic area *Photo Credit* Website of the People's Government of Xinxian County

4.4.2.3 Constantly Improving the Capability of "Liangshan" Construction by Strengthening the Innovation of Systems and Mechanisms on an Ongoing Basis

The Party Committee and the People's Government of Xinxian County have set up a directorate for the work of "Liangshan" practice and innovation bases, so as to reinforce the organization and coordination of construction work. The county has held a high-level and large-scale work conference on the construction of an ecological civilized demonstration county every year, formulated an annual implementation plan, clarified annual goals and tasks, signed target responsibility statements, and conducted a year-end assessment. It has implemented the "one vote veto system" for environmental protection projects, supervised and reported on construction work on a regular basis, and identified problems and urged the implementation of rectification in a timely manner. In addition, it has resolutely carried out the practice that "the Party and the government shall assume the same responsibility with one post having dual duties" in the ecological conservation arena, and successively mapped out a negative list for industry access in the key national ecological functional zone of Xinxian County, responsibilities of the Party Committee, the People's Government and relevant departments of Xinxian County for environmental protection work, measures on accountability of the Party and governmental leaders of Xinxian County for ecological and environmental damage (for trial implementation), *30 Rules for Forestry Resource Administration*, and other documents.

4.4.2.4 Consistently Raising People's Sense of Wellbeing and Gain by Promoting the Sharing of Ecological Benefits

Xinxian County is dedicated to improving the human settlement environment, strengthening urban management, solidly advancing the construction of itself as a national civilized city, and to conducting the comprehensive regulation of public order in urban areas. It has ranked first in Xinyang City in terms of urban management work for many consecutive years, and has been nominated as a national civilized city. The county has pushed forward the verdurization of urban areas, courtyards and suburbs, built a 5-km riverfront landscape belt, and constructed more than 20 leisure parks, so as to shape the "green heart of the city" and enable people to see and appreciate green sceneries both inside and outside their home. Xinxian County has actively implemented a five-year action plan on the construction of beautiful countryside and an ecological Xinxian County, and carried out the "double ten and double hundred project" for the construction of beautiful countryside and the "three cleansing and five rectification" activities, with a focus on controlling the phenomenon of "eight disorders". As a result, both sides of main trunks across the county have largely been kept "clean at the first glance", and market town locations, tourism scenic areas and spots, and administrative villages have achieved the goal of "three nos and one standard" (i.e. no garbage stacking, no sewage cross-flow, no debris blocking the way, and standardized and orderly stacking of necessities in daily production and life). The county has created 106 provincial- and municipal-level ecological villages, and all towns have become sanitary towns above the provincial level. For many years in a row, it has ranked among the top in Xinyang City and Henan Province in terms of improving the rural living environment. Xinxian County has constantly enlarged its ecological effect by building 189 industrial poverty alleviation demonstration bases. 75% of impoverished population and 130 leading enterprises and new types of agricultural businesses have established a close interest linkage mechanism.

Through share capital, employment salaries and asset rental, the annual average income of each household has increased by more than CNY3000. Xinxian County has continued to perfect the supportive policies for living security, and increase its support. In 2019, the poverty alleviation relocation of 73 poverty-stricken villages has been fully completed, making Xinxian County the first county in Xinyang City to be lifted out of poverty.

4.4.3 Key Achievements

4.4.3.1 Achievements in the Development of Ecological Industries

Xinxian County has built an ecological tourism mode of "all-area-advancing tourism and industrial integration". The whole county has been planned as a large scenic area, towns have been constructed as a scenic spot, and villages have been designed as a sketch, so as to form a trend of countywide development where "everywhere is brimming with scenery, every road represents a viewing gallery and every village has a tourist center", which has been summarized by the Xinhua News Agency as "there are sceneries but no scenic spots, there are scenic areas but no boundaries, and there are gates but no entrance ticket required". Two tourism brands of "Scenic Red City, Healthy Xinxian and Dabie Mountain Tourist Park" and "Travelling All Areas of Xinxian County Including 9 Towns and 18 Bays" have enjoyed a positive reputation nationwide. Xinxian County has completed more than 40 boutique projects of red tourism, with five scenic spots selected into the list of national classical red tourism scenic areas. Since the founding of the study tourism committee, Xinxian County has developed 24 boutique study routes, and built 39 red teaching points, in a bid to raise the popularity of the study tourism brand of "Dabie Red Camp". 29 red education and training institutions represented by the Dabieshan Executive Leadership Academy have provided training for 700,000 trainees in total, and the value of Dabie Mountain red education tourism brand has reached CNY3.862 billion. The county has cultivated 30 leading enterprises of agricultural industrialization at provincial and municipal levels, and developed 300,000 mu (or 20,000 ha) of tea, 280,000 mu (or 18,666.67 ha) of oil tea, 30,000 mu (or 2000 ha) of traditional Chinese medicinal materials, and 300,000 mu (or 20,000 ha) of timber forests with a gross forestry output value of CNY2.72 billion. Six featured agricultural products have obtained FDA certification, "Xinlin Yulu" has been appraised as a national well-known trademark, and "Xinxian General Dish" was granted the first geographical indication of agricultural products in Xinyang City. The county has actively promoted the transformation and upgrading of the tea industry, effectively driving 22,000 natives to increase their income steadily.

Flower sea in Zhouwan village *Photo Credit* Website of the People's Government of Xinxian County

4.4.3.2 Achievements in Ecological Poverty Alleviation

Xinxian County has built an ecological poverty alleviation mode of "cooperatives, ecological scenic areas and benefaction to people". Tianpu Dawan, Innovator Town and other typical cases have emerged. The county has persisted in not digging mountains, filling ponds, cutting down trees, or carrying out large-scale demolition and construction. Instead, it has formed a variety of formats including agritainment dining, characteristic homestays, and sightseeing experiences, thus realizing the transformation from beautiful villages to a "beautiful economy". Through the integration of agriculture and tourism, the county has generated invaluable assets relying on lucid waters. It has created three national key villages and 19 provincial-level distinctive villages for rural tourism, built 89 "Xinxian-style" series of agricultural product brands, and developed 340 types of tourism commodities. As a result, diverse experiences such as ecological sightseeing, fruit picking in the orchards, flower appreciation and photography have complemented one another. Xinxian County has retained its nostalgic atmosphere with quaint homestays, and developed 119 boutique homestays and 720 agritainment resorts, which have attracted numerous tourists. In 2018, the number of tourists in Xinxian County exceeded 6.3 million. The tourism industry has created more than 50,000 jobs and contributed 30% to the county's GDP. 22 poverty-stricken villages and 7683 locals in financial difficulties have eliminated poverty steadily through the development of tourism. Xinxian County has become the first region lifted out of poverty in the Dabie Mountain Revolutionary Old Area, and its poverty headcount ratio has declined from 14.9% to 1.12%, which has a typical demonstration effect.

Peach blossom on Hero mountain *Photo Credit* Website of the People's Government of Xinxian County

4.4.3.3 Achievements in Building Beautiful Villages

Xinxian County boasts a vegetation coverage rate of over 95%, a forest coverage rate of 76.7%, and an afforestation penetration rate of urban built-up areas of 100%. Its per capita green area of public spaces has reached 12.8 m^2, the annual number of days with good air quality has been kept at over 300, and the application amount of fertilizers and pesticides has witnessed zero growth. The county as a whole has implemented a three-year action plan on rural highways to promote urban and rural transportation integration projects. The county has built boutique tourism routes, and opened tourist buses connecting "9 towns and 18 bays". Consequently, the scenery lines mainly consisting of cultural corridors, ecological greenways, recreational walkways, and hiking trails have linked all villages across the county. The efficient and convenient intelligent platform of "Traveling around Xinxian" provides free Wi-Fi, 5G networks, time-sharing reservations, and voice-over narration to enhance the tourist experience. In addition, the number of service centers, tourism stations and ecological parking lots has grown significantly. Xinxian County has successively won several titles, including advanced unit for "toilet revolution" in the province, advanced county for improvement of rural living environment, and the first batch of cities with national secondary standards for ambient air quality.

4.4.3.4 Achievements in Rural Revitalization and Cultural Development

Xinxian County held the first national rural revitalization creativity competition for college students, sparking innovation and creativity amid lucid waters and lush mountains. The reform of the collective property rights system has been fully completed, the pilot reform of the home site system has been advanced in an orderly manner, and idle resources and assets have been activated at a fast pace. The county has implemented the "999" return project, set up the "Dabie Mountain Tourism Lecture Hall", and has constructed the first "China's Taobao Village" in Xinyang City. Furthermore, the county has pushed forward the "point-like land supply", developed the "fence economy", and promoted transformations from resources to assets, from capital to share capital, and from famers to shareholders. The county has explored a new mode of rural cultural cooperatives, launched new products for integration of culture and tourism, built civilization practice centers (offices and stations) in the new era covering county, town and village levels, and regularly hosts rural culture festivals.

Homestay alley in Tianpu dawan *Photo Credit* Website of the People's Government of Xinxian County

4.4.4 Inspirations

Over the years, following the development of all-area-advancing tourism in the old revolutionary base area and rural revitalization, Xinxian County has deeply implemented the concept of "Lucid Waters and Lush Mountains are Invaluable Assets". The county has promoted ecological conservation, made innovations in systems and mechanisms, and protected the ecological environment. The county has facilitated rural revitalization through green development, thus actively exploring a "red and green" road of ecological poverty alleviation. Based on its "red soil", the county has made "green accounts", played "green cards", and taken on "green paths". In doing so, the county has promoted the integrated development of cultural tourism, education and training, physical fitness, leisure wellness and other business forms, and driven the economic development and ecological poverty alleviation in the whole county, helping raise the green economy to new heights.

4.4.4.1 Adhering to a Strategy of Building an Ecological County to Consolidate Institutional and Mechanism Guarantees

Xinxian County has firmly established the concept of "red guidance and green development", giving top priority to ecological construction and green development in its economic and social development. Putting equal emphasis on concept establishment and institutional constraint, Xinxian County has advocated for the establishment of ecological consciousness throughout the area, laying a solid foundation for ecological environment protection among the public. Xinxian County has imposed hard constraints within various systems, and included about 43% of its land area into the protection scope of ecological red lines for plans to build Xinxian County into a national ecological conservation demonstration county. Furthermore, the county has carried out comprehensive control actions to crack down on four ecological "disorders", i.e. random excavation, disorderly construction, deforestation and unauthorized mining, making the ecological "red line" become a "live high-voltage line". The county has introduced a series of administrative methods and policy measures, such as measures on ecological conservation, ecological surveillance, and accountability for ecological and environmental damage. The county has also incorporated controlling indicators of ecological conservation and ecological civilization development into work report of the local government, and subjected them to the deliberation and approval of deputies to the local people's congress, so as to provide a strong guarantee for ecological civilization development in terms of leadership strength, systems and mechanisms, and objective management.

4.4.4.2 Putting Equal Emphasis on Construction and Management to Continuously Improve the Quality of the Ecological Environment

Xinxian County has insisted on ecological environment protection and restoration to strive to improve the quality of its ecological environment and its ability to supply ecological products. The county has implemented a strict ecological environment protection system. As for ecologically sensitive areas, eco-systematically important areas, and those playing a key role in natural reproduction of species, biodiversity protection, and regional ecological security maintenance, Xinxian County has delineated and strictly held an ecological red line and restricted the development and construction of red line areas for ecological conservation in a stringent manner, thus preserving original natural ecosystems to the greatest extent possible. The county has implemented forestry ecological improvement projects, strengthened biodiversity protection, and advanced ecological governance of mines. The county has also taken actions on prevention and control of air pollution and water pollution, and conducted soil pollution surveys. Through a series of measures for ecological environment protection and governance projects, the quality of the ecological environment in Xinxian County has improved steadily, and remarkable results have been achieved in ecological conservation.

4.4.4.3 Pushing Ahead with the Development of Ecological Industries to Allow Ecological Advantages to Consistently Unleash Green Momentum

Xinxian County has clarified key orientation for high-quality development, made great efforts in pollution prevention and control, and refined, optimized and strengthened distinctive ecological industries. In recent years, the county has strengthened ecological construction, reinforced protection of the Dabie Mountain ecological function area for water and soil conservation, and taken the lead in executing the "forest chief system" in Henan Province. The county has implemented a three-year action plan for land greening of "Beautiful Dabie Mountain · Green Mountains and Rivers" to build urban and rural greenways, landscape passages and ecological corridors. Ecological civilization development has become a new driver for green development of Xinxian County's economy while changing its appearance. Ecological tourism and ecological agriculture on the basis of a sound ecological environment have developed rapidly, providing a ceaseless ecological driving force for the green development of Xinxian County. The whole county has vigorously advanced practical explorations into transformation from lucid waters and lush mountains into invaluable assets, fully enhanced its ability to supply ecological products, actively developed green processing of agricultural products, and guided the development of green industry dominated by emerging sectors. The county has promoted the transition and upgrading of coal, salt and chemical industries as well as other traditional industries, and facilitated the integrated development of ecological industries for green environmental protection and sci-tech innovation, so that ecological industries can develop rapidly, and the vitality of ecological industries can be constantly enhanced.

4.4.4.4 Adopting Multiple Measures to Proactively Expand Methods for Ecological Poverty Alleviation

Making use of the profound red history of Dabie Mountain, Xinxian County has been nourished by the spirit of red revolution. Based on its advantages, Xinxian County has vigorously promoted ecological conservation, invigorated its rural economy through the "red and green" development mode, and realized a magnificent "turnaround" of the old revolutionary base area. Taking the "tourism+" concept as an engine, Xinxian County has constantly advanced deep integration of various industries, energetically developed all-area-advancing tourism, and extended industrial chains. The county has encouraged farmer "innovators" to explore the operation mode of "companies, cooperatives and farmer households", built innovator towns and developed characteristic homestays, cultivated and developed featured tourism products, and set up entrepreneurship demonstration bases. The county has organized a variety of cultural tourism activities, festival activities and literary and artistic creation activities around Dabie Mountain, build red education and training institutions, and held various sports events. The county has attached great importance to poverty alleviation through consumption, and expanded sales channels of featured agricultural products via internet platforms. The combination of a series of innovative measures has boosted rural revitalization in Xinxian County, pushing for the masses to achieve prosperity.

4.5 Fuliang County (Jiangxi Province)—Flourishing Ecology and Industries Boosting Rural Revitalization

4.5.1 Background

Located in the northeastern hilly and mountainous areas of Jiangxi Province and its border with Anhui Province, Fuliang County is advantageous in its location. With an area of 2851 km^2, it has long been known as "an area with 80% of its land taken up by mountains, 10% by farmland, 5% by water, and the rest taken up by roads and manors", which endows it with exceptional ecological advantages. Taking a prominent position ecologically, Fuliang County is the place of origin of the Raohe River (Changjiang River). As the most important ecological barrier and national key ecological function zone in the northeastern area of Jiangxi Province, it undertakes an important ecological barrier function across the whole river basin. It is also located in the Mount Huangshan-Huaiyu Mountains biodiversity priority conservation area, which is among China's 9 ecological sound areas and 35 biodiversity priority conservation areas. Reputed as the "Source of the Ceramic Capital of the World" and "Hometown of China's Famous Tea", Fuliang has been globally renowned for its ceramic and tea culture since ancient times.

Fuliang County has always been firmly committed to the development concept of "Lucid Waters and Lush Mountains are Invaluable Assets", and conscientiously implemented the instructions conveyed by the General Secretary Xi Jinping during his two survey trips to Jiangxi Province—"establishing a 'Jiangxi sample' for building a beautiful China through high-quality development". It takes the philosophy of a community of life composed of "mountains, rivers, forests, farmlands, lakes and grasslands and humanity" as its guiding principle, stays on the "ecological road" and plays the "ecology card" unswervingly, and spares no effort to do a good job in conserving and showcasing mountains and rivers and in coordinating "ecology, culture and development". Fuliang has been praised as a national ecological conservation demonstration county, national recreational agriculture and rural tourism demonstration county, national greening model county, national pollution-free tea production demonstration county and national tea standardization demonstration county, having blazed a new trail in the field of ecological conservation and high-quality development.

Fuliang's picturesque scenery (light show at the ancient county Yamun) *Photo Credit* Website of the People's Government of Fuliang County

4.5.2 Specific Practice

4.5.2.1 Conducting Ecological Environment Conservation in an All-Round Manner to Consolidate the Fundamentals of Lucid Waters and Lush Mountains

Fuliang County has made coordinated efforts to conserve and restore mountains, rivers, forests, farmlands, lakes and grasslands in a systematic way, and vigorously promoted the "natural forest protection project", "grain for green project", "Yangzi River protection forest project" and "public welfare forest protection project". It is the first county in Jiangxi to implement a logging ban policy with regard to natural broad-leaved forests and forests within 1000 m from the highway, cementing the foundation of ecological security. It has managed levels of water, gas, soil and waste pollution as well as "cooking fumes, barbecue fumes, garbage burning fumes, motor vehicle exhaust fumes, coal boiler waste gas, industrial waste gas, construction site fugitive dust, road fugitive dust, transportation fugitive dust, and stockyard fugitive dust" in a coordinated way. It has pushed ahead with special actions on the management of medium and small rivers and hilly ponds, as well as river cleaning and other comprehensive water environment improvement activities, enhancing the environmental quality all round. Taking "creation of a national civilized city and national sanitary city and double urban repairs" as well as beautiful countryside construction as an effective lever, Fuliang County has established the first integrated disposal mode of urban and rural garbage, under which the operation and maintenance services are purchased from the third party, realizing "disposal of garbage on a daily basis" and "no littering". Fuliang County has steadily pushed forward the construction of a sponge city and "cleaning of rural domestic refuse, rural water sources and water bodies, excrement from livestock breeding and other waste from agricultural production, as well as transformation of toilets in rural households and rural public toilets and change of the bad habits that impact the rural human settlement environment" in rural areas, and implemented the integrated improvement of the human settlement environment in urban and rural areas. Thanks to these endeavors, it has been praised as a national sanitary county, continuously consolidating its fundamentals in terms of lucid waters and lush mountains.

New village of Shiziyuan in Zangwan township *Photo Credit* Website of the People's Government of Fuliang County

4.5.2.2 Paving the Way for the Rise of "Fuliang Tea" and Reshaping Its Brand

Fuliang County has energetically carried out the "Fuliang Tea" brand strategy, remained committed to green ecological development, improved the quality and brand of the tea industry, and enhanced the core competitiveness of Fuliang Tea based on its ecological strengths, thus promoting the rise of Fuliang Tea. It has vigorously promoted the development of the tea industry, cultivated a green tea industry, and has built a green organic tea garden base of 300,000 mu (or 20,000 ha), pushing forward the popularity and improvement of the production technology of organic teas. With Fuliang Tea as its primary brand, Fuliang County has enhanced industrial quality based on the construction of organic tea bases and authentication of "pollution-free agricultural products, green food, organic food and agricultural products with geographical indications", thus pushing forward the in-depth development of the tea industry. It has innovated the operation mode of organic teas, and facilitated the transformation of agricultural production and operation modes by implementing a "companies + bases + rural households" operation mode through investment with land and capital, subjecting mountain forests and tea gardens to unified operation and management while realizing shared benefits for companies and rural households. The "Fuliang Tea" brand has made its way into the three brands with the highest brand resource power in the value assessment of regional public brands of Chinese tea, and was honored with a special gold award in the 4th Asia–Pacific Tea Competition. Among the first batch of brands to enter the China National Tea Museum, it has also been honored as a national geographic indication protection product and been awarded many global and domestic gold awards.

Ecological organic tea garden *Photo Credit* Website of the People's Government of Fuliang County

4.5.2.3 Fully Exploring the Connotations of Ceramic Culture to Light up the Brand of the "Source of the Ceramic Capital of the World"

Fuliang County boasts a profound ceramic cultural deposit. The Gaoling scenic area, located in Gaoling Village, Yaoli Town, Fuliang County, was the most important place of origin of raw materials of the ceramic manufacturing industry of Jingdezhen City in ancient times, and also the place where "Kaoline" (an international common special term of clay mineralogy) was named. Fuliang County has pushed forward the development of ecological industries related to ceramics by deeply exploring the profound connotations of its ceramic culture and fully combining its landscape advantages. It has actively and effectively protected and utilized Kaoline, mine pits, kiln sites and other ancient ceramic cultural relics, and fully demonstrated its brand through such projects as delineation of the natural reserve area in mines, implementation of mine environment governance and protection of Dongbu-Gaoling traditional villages. Taking the approach of "government guidance, enterprise participation and market-oriented operation", it has constructed the "Gaoling · China Village" project, in a bid to build a state-level pastoral complex integrating ceramic, tea and cultural tourism, ecological agriculture, forums and training, sports and health care, travel and residence, cultural innovation and other functions and commercial activities, uniting production, life and ecology as one. Relying on the R&D advantages of Jingdezhen Ceramic University and other colleges and universities, Fuliang County has made great efforts to foster new drivers of the ceramic industry, incubating the Sanlong High-grade Construction Ceramics Industrial Cluster and Xianghu Special Ceramics Industrial Cluster.

4.5.2.4 Promoting Integrated Development of "Green, Anciency and Red Culture" "in Trinity" to Create Health Care and Tourism Destination Brands

Leveraging its advantages in tourism resources with unique cultural and ecological characteristics such as "green, anciency and red culture" and its transportation location, Fuliang County has made continuous efforts to enrich its tourism topics and products, extended its industrial chain of tourism, and tapped into the overall function of tourism in activating overall development with the aim of "developing all-area-advancing ecological tourism and constructing eco-tourism resorts" and the construction of "two scenic areas and two renowned villages" as its key aim. Firstly, it has "developed based on momentum through introduction". Fuliang County successively introduced the Lifang Huaqiangu rural tourism project in Jiaotan Town with a total investment of CNY1 billion, Cangxi tourism development project invested by Jiangxi Changyun Co., Ltd., Hongqifeng eco-tourism project in Jinggongqiao Town with an investment of CNY300 million, a pastoral complex project with a total investment of CNY5 billion from Jingdezhen Coking Industry Group Co., Ltd. It has also "gone outside to publicize the brand". Fuliang County has used campaigns such as a bicycle competition around Poyang Lake, global "green shoe action" (large-scale hiking) platforms, carried out large-scale healthy hiking, bicycle travel and other activities, and held large-scale cultural publicity activities including drift festivals, tent festivals, a ceramic expo, tourism image ambassador qualification trials, and the "Fuliang Black Tea Hometown · Impressive Yantai" ancient tea garden opening event, in a bid to make the eco-tourism brand of Fuliang renowned and recognized by more people.

Traditional ancient village of Yantai village in Jiangcun township *Photo Credit* Website of the People's Government of Fuliang County

4.5.3 *Key Achievements*

4.5.3.1 Ecological Environment Quality Ranks Among the Top

As of 2019, forest coverage of Fuliang County had hit 81.4%, and its natural forest protection area amounted to 1.03 million mu (or 68,666.67 ha). Its ecological environment index (EI) remains at a superior level, ranking among the top in Jiangxi Province. It has been accredited as a provincial advanced county for afforestation project construction passageway greening and an advanced county for forest resource protection, and is a national pilot county for sustainable management of forests. In

2019, the number of days when the ambient air of Fuliang County met relevant standards was 337, taking a share of 92.3%, which meant its air quality remained acceptable continuously. The surface water environment quality remained acceptable at all times for three consecutive years, and the surface water quality and the water quality of centralized drinking water sources were rated as good overall.

4.5.3.2 Advantageous Ecological Resources Bring "People Affluence"

Building on its ecological resource advantages, Fuliang County has vigorously developed ecological industries and formed a development pattern which "takes the ecological industry as the basis, cultural industry as the core, health care industry as the support, and tourism industry as a lever", improving its comprehensive strengths at a steady pace. The GRP of Fuliang County rose from CNY10.01 billion in 2015 to CNY13.622 billion in 2019, representing an increase of 36%. The per capita disposable income of rural residents increased from CNY12,769 in 2015 to CNY18,014 in 2019, representing an increase of 41%. The per capita disposable income of urban residents grew from CNY23,674 in 2015 to CNY32,708 in 2019, representing an increase of 38%. Through implementing the "Fuliang Tea" brand strategy, Fuliang County was rated as the "National Charming Tea Hometown" and "Hometown of China's Famous Tea". The assessed value of "Fuliang Tea" brand in the regional public brand assessment system of national tea amounted to CNY2.376 billion. In 2019, the total area of the tea garden of Fuliang County reached 195,200 mu (or 13,013.32 ha), and the area under organic management standards was 60,000 mu (or 4000 ha), taking a share of up to 30.7%. The production value of tea skyrocketed from less than CNY100 per 500 g to over CNY500 per 500 g, reaching CNY735 million in the primary industry in 2019 and CNY1.7 billion in terms of overall production value. The team of tea growers expanded to 61,000 people, and tea-related per capita income amounted to CNY12,000. As a result of this, Fuliang County has initially realized its goal of both "affluent people and a beautiful ecological environment".

4.5.3.3 "Liangshan" Type Modernized Industrial System Takes Shape

Relying on tea, ceramics, forests, farmlands, mines, rivers, bamboos, schools, and other resource endowments, Fuliang County has adhered to an integration of primary, secondary and tertiary industries, as well as the integration of ecology, production and life. With its key focus on the eight industries, i.e. ceramics and tea, agriculture, forests, bamboo and rivers, advanced ceramic materials, green mining, cultural innovation, health care and tourism, universal aviation logistics, and the digital economy, Fuliang County has actively developed modern intelligent agriculture, new green industries and modern service industry, initially forming a "Liangshan" type modern industrial system with Fuliang characteristics. The industrial mix of the three industries was optimized from 13.0:59.6:27.4 in 2015 to 14.4:49.5:36.1 in 2019. The contribution of the service industry to the economy increased from 22.5 to 37.1%, and the proportion of the added value of the new- and high-tech industry in the GRP of the county rose to 7.5%.

Fuliang with picturesque scenery (light show at the ancient county Yamun) *Photo Credit* Website of the People's Government of Fuliang County

4.5.4 Inspirations

Fuliang County has combined learning with practice, taking the theory of "Lucid Waters and Lush Mountains are Invaluable Assets" as the guiding principle for driving high-quality development. With its natural advantages in ecological resources and location, Fuliang County has proactively explored effective ways to turn lucid waters and lush mountains into invaluable assets, making great efforts to develop its ecological industry and to transform ecological value into economic value. Fuliang County persists in taking historic culture as a starting point and developing based on the ecological space with characteristic industries as samples. It has pushed forward the organic integration of urban and rural resources, rural revitalization and green development, presenting a new landscape of Fuliang featuring "scenic rural houses and roads, as well as picturesque villages and towns". This mode of "integrated development of ecology and industries" for revitalization of Fuliang County applies to areas with abundant ecological resources. Utilizing its ecological superiority and taking ecology as its starting point, Fuliang County has actively investigated methods of turning ecological resources into humanistic resources and industrial resources and the mode of transforming lucid waters and lush mountains into invaluable assets, in a bid to enrich transformation channels and make transformed results sustainable. It has strived to create the "Liangshan" modernized industrial system, realizing integrated development of ecology, production and life.

4.6 Lishui City (Zhejiang Province)—Innovative Forest Rights Reform Promoting Trading in the Ecological Market

4.6.1 Background

Situated in the southwestern part of Zhejiang Province, Lishui City has a typical topographic feature of "90% of the area being mountains, 5% being water areas and the rest taken up by farmlands". Forests take up an area of about 21,991,900 mu (or 1,466,126.52 ha) and the forest coverage amounts to 81.7%, both ranking first in Zhejiang Province. As such, the area has long been reputed as "Zhejiang Green Valley" and "Forest Sea in Southern Zhejiang". There are also nearly ten thousand riverways in Lishui City, where numerous river systems intersect with one another. With natural advantages in ecological environment, climate conditions, and mountain resources, Lishui City has strengthened policy innovations, and conducted successful groundbreaking reforms such as the "forest rights reform" and the reform of "assigning the river rights to specific households", through which it has wakened up forests, riverways and other "assets in slumber" and accumulated advanced experience in ecological resource value transformation, all while making Lishui City a vivid financial practice sample of "Lucid waters and Lush Mountains are Invaluable Assets".

Longquan city's introduction of a policy benefiting the people *Photo Credit* Website of the People's Government of Lishui City

4.6.2 Specific Practice

4.6.2.1 Improving Supporting Policies and Institutional Platforms to Provide Basic Guarantees for Forest Rights Reform

Over the course of forest rights reform, Lishui City has established complete supporting policies and institutional platforms with regard to the difficulties faced by forest farmers in mortgage financing, a core issue that impedes the development of the forest industry. Lishui has increased the participation of financial institutions and provided various basic guarantees for forest rights reform. The city has formulated and unveiled a series of institutional documents and formed a complete institutional system covering forest rights confirmation certificate issuance, value assessment, mortgage registration, loan placement, transaction transfer, judicial disposal, risk mitigation, and financial incentives, among other aspects. It has constructed a platform system composed of "three centers and one institution"—the forest rights management center, forest resource collection and storage center, forest rights transaction center and forest resource asset investigation and assessment institution, with their functions covering forest rights assessment, mortgage registration, transaction transfer, mortgage guarantee, non-performing loan disposal, and other links and processes. It has also set up a standardized forest resource transfer and forest rights mortgage loan platform with high operation efficiency. It has taken further steps to promote the scope and volume expansion of forest-related financial services. All agriculture-related financial institutions in Lishui City have participated in and formed a financial service system regarding forest rights mortgage loans, which has also witnessed the extensive participation of state-owned commercial banks, joint-stock commercial banks, and urban commercial banks.

"Yongjundai" in Dazhangkeng village of Jingning county fueling rural development *Photo Credit* Website of the People's Government of Lishui City

4.6.2.2 Groundbreaking Innovations Have Been Launched to Boost Forest Rights Reform

Lishui has been exploring the forest rights system reform since 2006. On the basis of improving the policy environment and platform framework, it has launched small revolving loans for forest farmers, loans based on the direct mortgage of forest rights, loans based on the guarantee of forest resource collection and storage center, loans based on the mortgage of forest transfer and management rights, loans based on pledge of ecological public welfare forest compensation income right and other forest rights mortgage loan products of multiple varieties in an innovative way. This combines the practice of forest farmers, effectively meeting the capital needs of borrowers with various business types and different asset status and realizing inclusive finance to the maximum extent. In addition, during the reform, it has continued to optimize processes, remove institutional barriers and enhance the sense of fulfillment of forest farmers. In 2008, Qingyuan County of Lishui City launched China's first forest rights information system ("Forest Rights IC Card") featuring "unified assessment, one card for one household, and on-demand loans", which was then popularized in the whole city and even in the whole province. As the "electronic ID card" of forests, "Forest Rights IC Card" and "digital records" document the details of the forest rights owned by forest farmers and the assessed value of resource assets, which facilitates the application of loans of forest farmers, shortening the handling time of forest rights mortgage loans from one week to one day. Regarding issues such as the difficulty of new type forest business entities in guaranteeing their rights and interests, and in financing upon transfer of forests, Longquan City has taken the lead across China to divide the right of use of forests into forest contracting rights and forest management rights in 2013, issuing the *Forest Management Right Transfer Certificate* to business entities transferring forests, and coordinating with the financial department to carry out mortgage loans based on the *Forest Management Right Transfer Certificate*, in a bid to encourage the transfer of forest rights and promote industrialized management of the forest industry. In this way, the vision that "everyone that has mountain forests can benefit from forests though not all of them operate forests" has been realized. Although public welfare forests in Lishui feature a large area and steady compensation income with a high total amount, it's hard to dispose of forest assets which lack effective development and utilization. Therefore, many forest farmers struggled to conduct production and management. In such case, Lishui City has proactively innovated green finance, established the public welfare forest compensation income right certificate system, and created direct pledge loans, income guarantee fund loans, income right trust and other multiple kinds of public welfare forest compensation income right pledge financing modes based on the public welfare forest compensation income right ownership and compensation standards. In addition, relying on the expected income of the under-forest economy, Lishui City is the first in China to launch a pilot program for under-forest economy expected income loans, opening up a new way for broadening financing channels for forest farmers.

Village-enterprise cooperation for transforming lucid waters and lush mountains into invaluable assets to facilitate value realization of ecological products in Wuxi township of Yunhe county *Photo Credit* Website of the People's Government of Lishui City

4.6.2.3 Mutual Guarantee Sets up Security Lines for Forest Rights Reform

The fact that non-performing loans occurring to forest assets defy realization is a significant factor that makes financial institutions less willing to provide loans to forest farmers. For this reason, Lishui City has tapped into the potential of mutual assistance, cooperation and guarantee among forest farmers and explored an effective loan risk control mode, under which rural mutual assistance and guarantee organizations provide third-party guarantee for loans of forest farmers. Rural mutual assistance and guarantee organizations consist of self-owned funds of village collectives and specialized farmer cooperatives, funds of village-level share-based economic cooperatives, contributions by farmers to purchase shares, and public welfare forest compensation income funds, while farmers' loan guarantee funds have been set up. The organization provides guarantees to the members of cooperatives when they apply to financial institutions for borrowing, and farmers provide counter guarantee to the organization with their own related assets for the bank to provide a certain interest rate discount. When members of cooperatives need loans, cooperatives shall organize credit investigation, asset assessment and non-performing loan disposal and realization on their own. Through rural mutual assistance and guarantee, rural property rights such as rural houses, contracted land and forests are efficiently used, thus providing farmers with easier and more convenient access to loans.

4.6.2.4 Aligning Responsibilities with Duties, Providing New Drivers to Addressing River Governance Challenges

With a great number of riverways, Lishui City has long been challenged by the daunting tasks of riverway cleaning and management on very limited related budgets. Since 2014, on the basis of forest rights reform, Lishui City has been carrying out river rights reform with "assigning river rights to specific households" as its main approach. Such reform aligns the riverway operation income with river maintenance obligations, and establishes a long-term management mechanism featuring "conserving the river based on its own resources". The "river rights reform" in Lishui City contracts out the riverway management right in sections like contracting of farmlands and mountains. The contractor has the right to operate characteristic fishery businesses and rural recreational tourism businesses on the riverway, but also the responsibility to manage and protect the riverway on a routine basis. The mode of aligning responsibilities with duties turns passive river governance into active river governance, government governance into joint governance, and management weaknesses into long-term management samples, realizing the goal of "maintaining rivers based on their own resources". Through contracting, a market-oriented method, the government no longer needs to engage people to keep the riverway clean. The practice of contracting out the management right to villages, households and even specific persons arouses local people's wisdom and enthusiasm in entrepreneurship, which is helpful to exploit the potential value of riverway resources.

4.6.3 Key Achievements

4.6.3.1 Attaining Remarkable Achievements in the Fundamental Work of Right Decentralization and Confirmation

In forest rights reform and river rights reform, Lishui City has taken "right decentralization and confirmation" as significant fundamental work for establishing ecological product value realization mechanisms. For example, it has actively promoted natural resources right confirmation registration, and taken the lead in piloting the forest management right transfer certificate system in China, in a bid to confirm the forest management right transferred. It has pushed forward the reform of "assigning river rights to specific households" across the board and put in place the long-term management and maintenance mechanism featuring "conserving the river based on its own resources", among others. Only on the basis of clear confirmation of rights can corresponding rights and obligations be specified, local people's enthusiasm be better mobilized, and the foundation of green finance be more solid. At the end of December 2019, Lishui City had accumulatively completed the reform of "assigning river rights to specific households" with regard to 346 (or sections of) riverways, involving a total length of approximately 1520 km, 54 townships, and 234 administrative villages. The annual average increase in income per kilometer of the riverway registered over CNY8000.

4.6.3.2 Comprehensively Enhancing Information-Based Management

Lishui has established the first "Forest Rights IC Card" system in China, which stores the specific information of users with forest resources and simplifies loan assessment procedures. For common "scattered, small and plentiful" mountain forests, there is no need to assess them one by one on site, which saves human and other resources and facilitates the borrowing of money for forest farmers. The system standardizes forest rights records, while the forest rights and cadastral information with consistent "figures, tables and brochures" provides a foundation for the original informatization data of the forest industry, which is conducive to the formulation of forest management plans and forest logging management and resource protection. The system also safeguards the rights and interests of forest farmers, and solves the problem of unclear boundaries when determining forest rights in the case of hilly lands and forests, delineation of self-retained hilly land and identification of forest production accountability system, thereby minimizing the hazards of forest disputes. Finally, the system improves the credit environment as banks have basic information of the credit line and credit loans of forest farmers. At the end of August of 2020, Lishui City had totally issued 231,162 forest rights mortgage loans, with a total amount of CNY28.18 billion and the remaining loans being CNY6.534 billion.

4.6.3.3 Continuously Improving Risk Control Construction

Rural mutual assistance and guarantee organizations make use of the information advantage of a rural acquaintance society and fill in the gap of third-party intermediaries, resolving difficulties in assessment of agriculture-related assets and of forest farmers in obtaining guarantees. In addition, the procedures are simplified and financing costs are lowered, and the problem of handling forest assets upon occurrence of non-performing loans is avoided to a large extent for the bank, which leads to positive results for all.

"Joint governance of subdistricts and townships" opening a new chapter for community governance
Photo Credit Website of the People's Government of Lishui City

4.6.4 Inspirations

Lishui City has promoted collective forest rights system reform across the board and explored a market-oriented mechanism for ecological product value realization. It has established a forest rights management center, forest rights transaction center, forest resources assessment center, and forest rights mortgage loan service center, and built an ecological credit system and ecological product transaction system. On the basis of safeguarding the rights and interests of farmers, it has guided farmers in exploring new modes of forest rights transfer and scaled operations, and transferring forest rights through assignment, profit distribution, leases, trusteeship, stock investment and mutual exchange. It has witnessed the formation of family forest type, trusteeship forest nurse types, share-based cooperation types, lease-based management types, specialized cooperation types and other multiple new types of market-based forest business entities, and addressed the three challenges in effectively using forest resource assets, applying for loans with forest rights as the collateral, and issuing management right certificates. It has awakened the "assets in slumber" in the mountains, and effectively conducted risk control and right decentralization and conformation, as well as other fundamental work based on actual local conditions through comprehensively improving the information-based management level, which has fully motivated local people in joint construction, co-governance and sharing, turned the beautiful ecological environment into an "economic factor" and solidified the foundations of green finance. The new forest rights reform of Lishui City has specified, stabilized and validated the responsibilities, rights and interests of forest rights, realized the "determination of boundary of hilly land and the rights of forests, and reassurance of people", "transformation of live trees into money" and "transformation of ecological resources into assets". It has provided a lot of valuable experience that can be referred to for other areas in China in fully tapping into the potential value of ecological resources, utilizing the allocation function of the market-oriented economy, and striving to open up the channel of turning lucid waters and lush mountains into invaluable assets for other areas in China.

4.7 Anji County (Zhejiang Province)—"Liangshan Bank" Propelling the Transformation of Ecological Resources into Assets and Capital

4.7.1 Background

Located in the northwest of Zhejiang Province and at the geographic center of the Yangtze River Delta, Anji County is the source of the Huangpu River and an important northwestern node of the Hangzhou metropolitan circle, covering an area of 1886km^2. 70% of its area is taken up by mountains, 20% by farmlands and 10% by water areas. With a mountainous landscape where bamboo stretches across the view, Anji is reputed as "an area with clean air, water and soil". Anji County is in the northern subzone of the mid-subtropical evergreen broad-leaved forest, with a great variety of both southern and northern biological species. Its vegetation coverage and forest coverage reach 75% and 70.1%, respectively. In 2020, the GRP of the whole county amounted to CNY48.706 billion, and its total fiscal income exceeded CNY10 billion for the first time. As the birthplace of General Secretary Xi Jinping's concept of "Lucid Waters and Lush Mountains are Invaluable Assets" and the birthplace of beautiful villages of China, it is among the first batch of "Lucid Waters and Lush Mountains are Invaluable Assets" practice and innovation bases, the second batch of national ecological conservation demonstration counties, and first batch of counties obtaining the National Ecological Civilization Award. It is also the first county in China to obtain the "UN Habitat Scroll of Honor Award".

Lucid waters and lush mountains are invaluable assets *Photo Credit* Website of the People's Government of Anji County

Relying on unique resource endowments and location conditions, Anji County has been practicing the theory of "Lucid Waters and Lush Mountains Are Invaluable Assets" for many years. It has focused on reform and innovation and explored a path of green development featuring beautiful ecology, prosperous industries and affluent people. In 2020, in order to make the transformation of "lucid waters and lush mountains into invaluable assets" more intensified and brand-oriented, and break the "bottleneck" that restricts the realization of high-level transformation of "lucid waters and lush mountains into invaluable assets", the Party committee and government of Anji County took the lead in China to issue the "Liangshan Bank" pilot implementation plan after in-depth research, actively launching the pilot program in a bid to "deposit" lucid waters and lush mountains to "withdraw" invaluable assets. By referring to the decentralized input and overall output mode of commercial banks, Anji County carried out ecological resource storage transactions under a unified mode across the whole county through building an ecological resource operation service system featuring government guidance, market-oriented operation, and participation by people from all walks of life, further improved the ecological resource value, accelerated the construction of the mechanism for high-level realization of ecological product value, and enriched the pathways of turning ecological resources into assets and capital.

4.7.2 Specific Practice

4.7.2.1 Constructing the System Operation System of "Liangshan Bank" Under Government Leadership

Anji County has set up a county-level ecological resource asset operation and management platform company (hereinafter referred to as "Liangshan" Company)

mainly based on Anji City Construction Investment Co., Ltd., a state-owned company in Anji, which is composed of the Collection and Storage Department, Data Department, Assessment Department, Credit Department, Transaction Department, Risk Control Department and other departments. Based on rights confirmation registration and the reform of "separating ownership right, contract right, and management right" regarding natural resources, these departments exercise property rights collection and storage and resource improvement, resource measurement and dynamic management, resource value assessment, resource project credit enhancement, resource packaging and market trading, whole-process risk control and other functions.

4.7.2.2 Ascertaining Resource Assets to Form a List

By means of remote sensing of satellites, blockchain and other digital methods, Anji County has conducted resource surveys, fully identified ecological fundamentals, rationally assessed the value of ecological resource assets, increased the pace of resource asset right confirmation registration and formed a list of ecological resources and a list of property rights through data integration, and put the data into the "Liangshan Bank" platform, thus enhancing monitoring and management of natural resource assets.

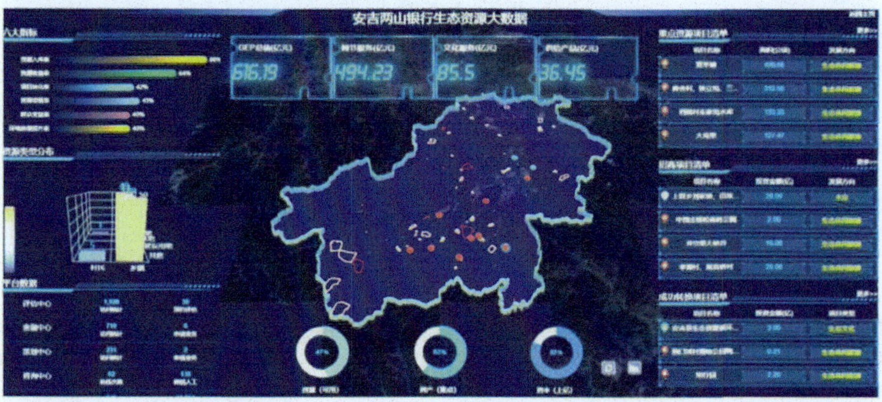

"Liangshan bank" ecological resource big data platform of Anji *Photo Credit* Publicity Department of the Party Committee of Anji County

4.7.2.3 Integrating and Improving Ecological Resources by Forming Project Packages of Natural Resource Assets

Anji County has collected and stored idle resources and inefficiently developed projects through leases, stock investment, trusteeship, redemption and other methods, transferring resources to the "Liangshan Bank" platform. Such scattered resources have been integrated and improved by "Liangshan" Company or other project companies in Anji County, forming concentrated, contiguous and high-quality natural resource asset project packages. Meanwhile, Anji County has established a project library to attract and introduce projects, and developed modern agriculture, rural tourism, health care, cultural creativity, boutique homestays and other new industries and formats through mixed ownership, stock cooperation, entrusted management and other methods, thus improving the benefits of comprehensive utilization of resources.

"Glory · castle in the sky" scenic area *Photo Credit* Publicity Department of the Party Committee of Anji County

4.7.2.4 Improving the Service System and Exploring Mechanism Innovation

Anji County has focused on difficult points and choke points in the process of propelling resources to convert into assets and capital. The county has made efforts to improve the consultation service mechanism, project attraction and introduction mechanism, and financial guarantee mechanism, and strengthened the support of its specialized consultation service. It has perfected the system for purchasing technical consultation service, established an expert committee, and introduced high-caliber professional consultation institutions, in a bid to provide technical support for the capitalization and industrialization of ecological assets. It has energetically introduced high-quality industrial investment institutions and excellent operation and management teams for cooperation, enhanced financial guarantee service, and propelled the cooperation with financial institutions to establish green industry, rural revitalization and other development funds which first support the project of realizing the value of ecological products.

4.7.2.5 Integrating and Advancing Reform Pilot Programs in a Steady and Orderly Manner

On the basis of the asset list, Anji County has selected Shangshu, Tianhuangping, Xiaofeng, Baofu, Zhangcun, Kenggai, Meixi and other townships as pilot areas after in-depth and concentrated surveys, and obtained a series of typical experience and system results in terms of the assessment, transfer, trading, operation and management of ecological resources. In the meantime, based on actual conditions, it has established the dynamic pilot adjustment mechanism, and included the work of establishing "Liangshan Bank" of pilot townships into appraisal. It has put focus on mountains, rivers, forests, farmlands, lakes, grasslands and other ecological resources, and proactively explored the reform of "separating the ownership right, contract right, and management right" regarding idle rural homesteads and idle rural houses, improved the supporting system for rural collective for-profit construction land to enter into the market, and comprehensively promoted provincial-level pilot programs for ecological restoration of mountains, rivers, forests, farmlands, lakes, grasslands and the pilot work of the mechanism for realizing the value of ecological products.

Idle rooftop turning into "sunshine passbook" *Photo Credit* Website of the People's Government of Anji County

4.7.2.6 Exploring the Mechanism of Co-Governance and Benefit Sharing Based on Win–win Cooperation

Anji County has explored a governance mode with the joint participation of the government, collectives and villagers, and established a complete hierarchical and classified resource asset management system. It has built the "Liangshan Bank" in county and township levels, based on which financial institutions provide capital support, bank operation modes and resource asset management and release systems and other factor guarantees to township banks. Meanwhile, it has improved the "enterprise, collective, cooperative and villager" and other business modes, encouraging rural collective economic organizations to invest in the stock of the project company and forming a mechanism of sharing benefits among enterprises, collectives and villagers. The all-win community based on four-level linkages has taken shape, which features "civilians' revenue increase through entrepreneurship, village collectives' business expansion, investors' harvest of revenues and profits and township government's strength improvement".

4.7.3 Key Achievements

4.7.3.1 Building a Platform for Transforming Resources to Assets and Capital

By creating a natural resource operation and management platform, "Liangshan Bank" integrates, transforms and improves ecological resources scattered across the county, and introduces social capital and specialized operators, thus turning resources into assets and capital and providing a basis and channel to realize the value of ecological products. Through establishing an ownership management mechanism of "right confirmation (investigation)—right decentralization (certificate issuance)—right use (guarantee)—right delegation (credit)—right activation (trading)", it pushes forward the trading of ecological resource rights and interests and continuously improves the ecological product operation and development mechanism. Currently, there are 108 key reserve projects for transformation of ecological resources across the whole county in the project library of Anji County, with a total planned investment of over CNY20 billion and involving existing construction land of about 2000 mu (or 133.33 ha), collective for-profit construction land of about 5000 mu (or 333.33 ha), forests of over 100,000 mu (or 6666.67 ha), water areas of approximately 1 km^2, and more than 200 idle rural houses. 24 "Liangshan" projects, which have attracted unified investment, have secured a total investment of CNY20 billion and effectively used 262 mu (or 17.47 ha) of existing idle land.

4.7.3.2 Improving the Value of Resources and Ecological Products

Through "Liangshan Bank", Anji County has managed and maintained its land, improved soil fertility and grown crops. It has done so with regard to the accepted projects of land reclamation for cultivation and those in previous years, ensuring that these projects deliver the greatest benefits effectively and stably in the long term. The first batch of 23 waste land reclamation and renovation demonstration villages in which the construction advancement is prioritized involves 7 townships (sub-districts). An area of land of 5210 mu (or 347.33 ha) has been ploughed, while the cultivated area transferred by the village collectives amounts to 2783 mu (or 185.53 ha). Through building a public brand system with regional features dominated by core brands such as "'Liangshan' Agricultural Products Collection" and "Anji White Tea", Anji County has strengthened its supervision of the quality of ecological products and enabled "good ecology to turn into good products and good products to propel the development of good brands", thus contributing to the value increase of ecological products.

4.7.3.3 Job Creation and Income Increase for Farmers

Anji County has ensured that rural households share the dividends of transforming "lucid waters and lush mountains to invaluable assets" through establishing a mechanism of income sharing among "Liangshan Bank", village collectives and rural households. "Liangshan Bank" has conducted cooperation with Shangshu Village and Yuanlin Village of Huashu Township, and collected and stored 302.42 mu (or 20.16 ha) and 429.7 mu (or 28.65 ha) of cultivated land. It has committed that the annual dividend per mu will be no less than 225 kg of unhusked rice (or CNY585). Since the establishment of "Liangshan Bank", 19 projects related to culture-tourism integration and effective use of idle resources have been successfully transformed, involving 10 townships (sub-districts) and 19 villages (communities). They have registered an operating income of CNY225 million, increased the village collectives' economic income by over CNY11 million, and created more than 1,430 jobs for local people. Meanwhile, 18 projects related to culture-tourism integration, effective use of idle resources, and the circular economy have been promoted. It's expected that such projects, after being put into operation, will generate an annual operating income of over CNY300 million, increase the village collectives' economic income by more than CNY14 million, and create over 1600 jobs for the general public.

Aquaculture base in Xiaoxikou village of Meixi town *Photo Credit* Website of the People's Government of Anji County

4.7.3.4 Opening up the Channel for Realization of the Value of Ecological Products

Based on its resource endowments and locational conditions, Anji County has introduced the "Turn Right in the Bamboo Grove" project, "Foot of Tianmu Mountains · Parent–Child Research Study Village" project, "Anji Hongruitian Chinese Peony Valley" project and "Intelligent and Beautiful Pasture" project. By "attracting, introducing and promoting projects", Anji County has consolidated factor guarantees for the reform pilot programs of "Liangshan Bank" on a continuous basis. Since the operation of "Liangshan Bank", the value of ecological products in the county has doubled, and the transformation of the value of ecological resources in the county has seen significant achievements. In 2020, the total village-level collective economic income of the whole county increased by 13.1% year-on-year, demonstrating that it has vividly interpreted the green development practice of "depositing lucid waters and lush mountains and withdrawing invaluable assets".

4.7.4 Inspirations

"Liangshan Bank" in Anji County has extended the concept of small savings for lump-sum withdrawal into the ecological asset resource development field, in reference to the operation mode of commercial banks. It has collected and stored fragmented ecological resources on a large scale, integrated them in a specialized way and operated them on a market-oriented basis. It has transformed ecological resources into high-quality asset packages, thus realizing the transformation of "lucid waters and lush mountains into invaluable assets" in a formal and efficient way, while retaining a high quality. "Liangshan Bank" has exerted itself to gather, transfer and improve the value of resources while leveraging the functions of the platform, thus positioning itself as having "ecological resource management, ecological resource investment attraction, industrial integrated development, ecological resource trading, financing guarantee, and ecological restoration" functions. Anji County has innovatively explored reform pilot programs of "Liangshan Bank", propelled ecological resources to be transformed into assets and capital, and opened up channels transforming "lucid waters and lush mountains" into "invaluable assets". This offers a transformation path and demonstrative sample for facilitating comprehensive rural revitalization and better achieving common prosperity.

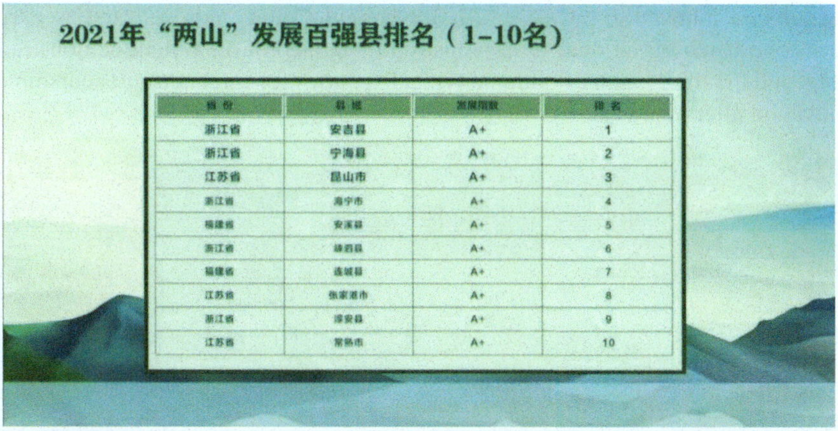

Anji county ranking first among the national top 100 counties in the development of “Liangshan” for four consecutive years *Photo Credit* Website of the People’s Government of Anji County

4.8 Chongyi County (Jiangxi Province)—Promoting Green Finance Reform to Facilitate the Transformation of “Lucid Waters and Lush Mountains into Invaluable Assets”

4.8.1 Background

Located at the southwestern edge of Jiangxi Province, Chongyi County covers a total area of 2206.27 km^2, with 6 towns and 10 townships under its jurisdiction. It is in the subtropical monsoon humid area and the evergreen broad-leaved forest climate zone. With a mountainous landscape, it has a terrain inclining from southwest to northeast. There are 83 rivers distributed across the county, both big and small. It is known as the “Green Treasure House of Jiangxi Province” as it boasts abundant forest resources. Chongyi County has long been reputed as “a place with 90% of its land being mountains, 5% being farmlands and the rest taken up by roads, water area and manors”. As a key national forest county in the southern collective forest region of China, it has won the titles of national key forest county, national greening model county, national forest city, national garden county, national sanitary county, etc., successively. It is the demonstrative unit in implementing the *UN Forest Instrument*, and is in the second batch of national ecological conservation demonstration counties and the third batch of “Lucid Waters and Lush Mountains are Invaluable Assets” practice and innovation bases. In 2020, Chongyi County was selected as part of the second batch of national rural industrial integrated development demonstration parks, and the third batch of national typical counties for rural entrepreneurship and innovation after appraisal.

Hakka terraced fields in Chongyi county of Zhejiang province *Photo Credit* Chongyi Ecological Environment Bureau

In recent years, embracing the concept of "Lucid Waters and Lush Mountains are Invaluable Assets", Chongyi County has proactively explored an effective approach to green finance by closely combining its ecological superiority with unique environmental resources, formulating an implementation plan for green finance reform pilot work of Chongyi County. It has probed into the establishment of standard systems, green enterprise libraries, and green project libraries, and implemented dynamic management to promote the construction of green projects. It has expanded green credit placement, set up the forest rights collection and storage center of Chongyi County, publicized a forest rights mortgage and pledge loan mode, innovated green credit products for characteristic industries, put in place a green loan risk compensation mechanism, strengthened refinancing support, and simplified the review and approval process of green credits, thus making green finance better serve green industries. The vivid practice of green finance reform in Chongyi County has opened up a channel for bidirectional conversion between "lucid waters and lush mountains" and "invaluable assets", better connected and integrated ecological resource advantages and capital advantages, effectively propelled the conversion of green ecological advantages into economic development advantages, and boosted high-quality economic and social development and ecological conservation of the county, blazing a new path of green finance fueling high-quality development and creating a "Chongyi sample" for green finance reform in counties.

4.8.2 Specific Practice

4.8.2.1 Strengthening Top-Level Design and Establishing a Complete Green Finance Development Plan

In September 2020, Chongyi County issued the *Implementation Plan for Green Finance Reform Pilot Work of Chongyi County*, which is the only county-level implementation plan for green finance reform pilot work in Jiangxi Province. It also developed a county-level green finance standard system, and introduced the *Methods for Identification of Green Enterprises of Chongyi County*, the *Methods for Identification of Green Project of Chongyi County* and other documents, which further clarifies how to construct a green finance system and how to give policy support, and forms a multi-layer green finance organization system and a multi-level service support system, laying a solid foundation for promoting green finance reform within the county.

4.8.2.2 Optimizing the Guarantee Mechanism and Establishing a Leading Group for Green Finance Reform Pilot Work

Chongyi County has set up a dedicated green finance service office to advance the green finance reform pilot work of the whole country in a coordinated way. The Financial Bureau of the county annually allocates a special fund of CNY2 million to support green finance development through coordination. Green finance has been included in the high-quality development appraisal and evaluation system of the county, and financial rewards will be given to financial institutions meeting targets or completing tasks in this area, thus promoting the implementation of green finance reform.

4.8.2.3 Optimizing Enterprise Financing Platforms to Promote the Operation of Ecological Resource Assets

Chongyi County has innovatively established an ecological branch of the Bank of China to build an ecological resource asset operation institution. Relying on equity market service platform resources of the green ecology category of Chongyi County, the county has introduced the *Implementation Opinions on Speeding up the Enterprise Shareholding Reform and Listing of Chongyi County*, established a green enterprise information sharing mechanism, and greatly supported green enterprises in the county in shareholding reform and listing. Lanyang Fuyuan, Tianrun Agriculture and other green enterprises have completed shareholding reform, and the two green enterprises of Qiyunshan Food and Lidao New Energy have started their shareholding reform process.

4.8.2.4 Innovatively Developing Green Insurance to Promote Effective Transformation of Resource Advantages

With its spine grapes, high mountain tea and other characteristic industries, Chongyi County has innovatively developed index-based weather insurance, yield insurance, price index insurance, income index insurance and other unique insurance products for characteristic agricultural products, and optimized county-level fiscal compensation standards synchronously. It has launched pilot programs on mandatory environmental pollution liability insurance and live pig "insurance and futures", innovating commercial insurance regarding ecology and environmental protection in a bid to propel the conversion of resource superiority into development strengths. It has accumulatively provided a risk deposit of CNY27 million to Shangbao Terraced Fields, Yangmingshan National Forest Park and other core scenic areas.

4.8.2.5 Broadening the Channel of Green Credits to Drive Green Economic Development

Chongyi County has created green bonds, green funds and other financial services, innovated the "government—enterprise—bank" connection and cooperation mechanism, established the "Green Industry Integration Xindaitong", established a risk deposit of CNY50 million jointly with enterprises based on the mode of "the government + leading enterprises", and obtained bank credit funds of CNY400 million, providing an all-round and professional financial credit service for green financing projects.

4.8.3 Key Achievements

4.8.3.1 Green Finance Accelerates Industry Integration

Chongyi County has brought into full play the agglomeration effect of a demonstration park of the integrated development of three industries, innovatively established a green finance service center, and successfully acquired loans with a low interest of USD24 million from Asian Development Bank and USD30 million from the World Bank for investment in green industries and projects. With the use of "Xindaitong" and other green financial products, it has set up Chongtianran Company, promoting the integrated development of such characteristic industries as Chongyi dumplings, high mountain tea and spine grapes. In 2020, Jiangxi Zhangjiangyuan Tourism Development Co., Ltd. issued special green project income corporate bonds of CNY500 million. Chongyi County Water Group carried out ABS (Asset Backed Securitization) financing of CNY200 million, realizing a green credit balance of CNY3.2 billion. This accounted for 30% of the loan balance of the whole county, representing an increase of 73% compared with the corresponding period of the previous year. Currently one green enterprise has been listed, 9 have completed shareholding reform, 6 have been listed at the standard level, and 24 have been displayed at the display level.

4.8.3.2 Green Finance Boosts the Development of the Forest Industry

Following approval of the Implementation Plan for Green Finance Reform Pilot Work of Chongyi County, Chongyi County has unveiled the Implementation Plan for Pilot Reform of the Paid Use System for State-owned Forest Resources Assets of Chongyi County, Measures for Paid Use of State-owned Forest Resources Assets of Chongyi County (for Trail Implementation), Implementation Plan for Green Finance to Boost the Development of the Forest Industry of Chongyi County and other supporting polices based on the actual work conditions of the forest industry. Chongyi has taken the lead to set up "Liangshan Bank" in the whole city, established a forest rights collection and storage service center, and opened a "one-stop" service window for forest rights collection and storage, laying a foundation for the capitalization of the forest right assets of the forest industry. It has innovated the forest rights mortgage guarantee form by adopting the method of joint contribution of "the government and associations". It has set up a forest rights collection and storage deposit of CNY50 million and deposited it into the bank as a loan risk deposit, based on which no less than 8 times' loans have been obtained from the partner bank for production activities of the forest industry. This has accelerated the process of turning resources into capital and capital into funds, effectively using "assets in slumber" of the forest industry across the board, and validly settled the problem of "insufficient endogenous power of banking institutions issuing loans to forest-related industries".

Chishuixian organic tea garden *Photo Credit* Chongyi Ecological Environment Bureau

4.8.3.3 Green Finance Empowers Brand Value Increase

With the support of green finance, Chongyi County has introduced green crowdfunding, regional co-construction and other innovative forms in developing ecological brands and built a "mother–child" brand system combining "1 + N" regional public and enterprise brands, further promoting effective improvement of the brand value of ecological products. Product sales of the latest-released regional brand "Chongshui Shantian" have topped CNY10 million in only a month. In 2020, the high mountain tea in Chongyi was awarded the Gold Award at the China Brands & Trademarks Exposition; Shangbao rice won the Gold Award for products at the Second Jiangxi Exposition themed at "Ecological Poyang Lake • Green Agricultural Products"; the two brands of Shangbao Xuezhuchun Green Tea and Houseone Ecological Board were praised as renowned products of Jiangxi.

4.8.3.4 Green Finance Has Improved Green Ecology

With the support of green finance, Chongyi County has implemented major biodiversity protection projects to improve forest quality in a targeted manner, successfully making its way into the first batch of county-level national forest cities of China and becoming a national forest health care base pilot construction county based on its advantageous resources. At present, Chongyi has built-up commercial timber bases of 1.4 million mu (or 93,333.33 ha), forest management sample demonstration forest bases of 510,000 mu (or 40,000 ha), 1 national nature reserve, 2 provincial-level nature reserves, and 1 national wetland park. It has successfully created 10 provincial-level ecological townships, 10 provincial-level ecological villages, and 83 city-level ecological villages, and the number of ecological villages above the city level accounts for 75% of the total number of administrative villages in the whole county.

Collection of some wild fruits in the wild fruit protection area of the Gentleman valley *Photo Credit* Chongyi Ecological Environment Bureau

4.8.4 *Inspirations*

Based on its rich resources, Chongyi County has proactively explored new ideas, methods, and mechanisms and optimized policies, resources, capital and other factor supplies, in a bid to fully consolidate the foundation of green finance development. It has channeled funds towards the real economy by promoting green finance product and service innovation and increasing the financing channels of green enterprises, which has comprehensively improved the endogenous power of real economy development and enables ecological resources to turn into assets. Contributing over 60% to the economy of the county, the green ecological economy has provided strong support to high-quality economic development of the county, which has realized the utilization of "invaluable assets" to reciprocate "lucid waters and lush mountains" and created a "Chongyi sample" for green finance reform in counties. The green finance reform and development mode of Chongyi County applies to areas with good ecological resource endowments, solid industrial foundations, and a high concentration of innovative factors, which facilitates coordinated development of industries and brand value improvement with green finance and builds a "green engine" for high-quality development of the county.

4.9 Quzhou City (Zhejiang Province)—Leading Green Finance Reform to Blaze a Path of Green Development

4.9.1 *Background*

Located at the Yangtze River Delta region, western part of Zhejiang Province, and west end of the Jinhua-Quzhou Basin, Quzhou City is known as a place at the "junction of four provinces and the central area of five administrative areas". Quzhou is an ecologically beautiful city famous for its mountains and rivers. Located at the source of the Qiantang River, it has a forest coverage of 71.5%, and is an important ecological barrier in Zhejiang Province and in East China as a whole. However, Quzhou City is also a traditional heavy chemical industry base, forming a heavy chemical industrial structure dominated by industry sectors of chemical, machinery, cement, and papermaking. The proportion of heavy industry is 68%, and the contradiction between the heavy chemical industrial structure and ecological conservation has become increasingly obvious. In order to protect the good ecological environment and complete the political tasks assigned by the provincial Party committee and provincial government of "ensuring the delivery of a clean river to Hangzhou", the city urgently needs to develop green finance, promote transformation and upgrading of traditional industries and boost ecological conservation.

Since 2017, when Quzhou was approved to be among the first batch of green finance reform innovation pilot zones in China under the guidance and push of the People's Bank of China, financial institutions in Quzhou have actively carried out regional green finance reform innovation practice, and built the "five pillars" of green finance with hard work based on local conditions and the "green and feature" philosophy, effectively promoting the green transition and sustainable development of local society and economy and accumulating valuable experience.

Demonstration section of "Quzhou, a model city of virtue", a picturesque scenic zone in Kecheng district *Photo Credit* Bureau of Agriculture and Rural Affairs of Quzhou City

4.9.2 Specific Practice

4.9.2.1 Establishing a Green Finance Standard System and Creating a "Quzhou Sample"

According to the *Plan for Development of Green Finance Standards of Zhejiang Province (2017–2022)*, Quzhou City has taken the lead in developing local standards for evaluation of green enterprises (projects) with regard to five major traditional industry sectors including chemical and papermaking and traditional agriculture such as the animal husbandry industry, and issued the *Standards for Evaluation of Green Enterprises* and *Standards for Evaluation of Green Projects*. The evaluation results have better validated the applicability of the standards and the leading role of such standards in green transition. In addition, Quzhou City has made continuous efforts to push forward green transformation of financial institutions, included "one-vote veto of the environmental protection department" across the whole credit management process, and publicized a green management model featuring "single presentation of credit scale, single presentation of capital price, single presentation of risk management indicators and single presentation of credit review and approval channels, single presentation of performance appraisal and single presentation of green financial products" and "exclusive institutions, special scale, special review and approval and special assessment".

4.9.2.2 Green Finance Products Are Enriched to Help Traditional Industries Pursue Green Transformation and Transition

The city of Quzhou has innovated pollution discharge right pledge loans, green energy conservation loans, technical upgrade loans, contract energy management loans, natural gas distributed energy project loans, "golden roof" photovoltaic loans and other green credit products, making financing more accessible to micro and small environmental and tech enterprises. It has introduced innovative "comprehensive liability insurance for safe production and environmental pollution", so as to support the transformation and upgrading of traditional industries. It has created comprehensive liability insurance for safe production and environmental pollution featuring "insurance + service", created live pig blanket insurance and comprehensive liability insurance for pollution-free disposal, safe production and environmental pollution and other green insurance products, and shared safety and environmental protection pressures through the utilization of commercial insurance mechanisms. Quzhou has innovated whole-chain financial services, supported green transition in the pig farming industry, and formed a type of modern animal husbandry industry development featuring "breeding at a standardized scale, comprehensive utilization of waste materials, continuous improvement of the environment and optimization of the industrial layout". The city has explored and innovated the green option concept of "one village with thousands of trees", under which local investors purchase asset packages from village collectives, and enjoy asset disposal rights upon expiry of the agreed term. A social force in purchasing the future income of forests and village collectives obtaining green capital funds has been formed, effectively opening up channels among resources, assets and capital of "one village with thousands of trees".

Orange town in Kecheng district *Photo Credit* Bureau of Agriculture and Rural Affairs of Quzhou City

4.9.2.3 The Carbon Account System Has Been Originated and Green and Low-Carbon Development Has Been Propelled Through Green Finance

Quzhou City has explored and set up enterprise carbon accounts and low-carbon evaluation standards based on "green energy code" data from the energy big data center, labelled enterprises with "red, yellow, light green and dark green" tags to classify their carbon emission levels, matched enterprises with differentiated fiscal and financial policies and green finance service, motivated traditional enterprises to go through green transformation and transition, and actively propelled the "Qurongtong" financial service platform to provide convenience to financial institutions in their identification of green and low-carbon enterprises and innovation of low-carbon financial products and service. It has established agricultural carbon neutrality accounts by taking traditional planting and breeding, recycling of the animal husbandry industry and fertilizer utilization as critical points of emission reduction, and given differentiated policy support to agricultural businesses with their agricultural carbon neutrality accounts rated at "light green, medium green and dark green" levels, so as to address the challenge of limited emission reduction space of agriculture. In addition to launching the first "personal carbon account" system in the bank in China, it has taken further steps to construct personal carbon accounts, and developed payment operations for 18 daily scenarios. Clients can exchange green articles for daily use, collect organic vegetables, and enjoy deduction upon consumption at cooperation merchants and other rights and interests offline through "personal carbon account credits", or convert such credits into public welfare donations to special groups in distress, thus guiding residents towards green life and consumption.

4.9.2.4 Comprehensively Promoting "Liangshan Bank" Reform Pilot Work and Exploring the Establishment of the Green Finance Mechanism

In 2020, as a national green finance reform innovation pilot zone, Quzhou became one of the two pioneering pilot cities of "Liangshan Bank" reform in the whole province, and Changshan County, Kaihua County and Qujiang District launched pilot work in this area. In order to practically conduct the "Liangshan Bank" reform pilot work, Qujiang District formulated the *Implementation Plan for "Liangshan Bank" Reform Pilot Work of Qujiang District*, which indicates that it will explore the establishment of a green finance mechanism for ecological products through running trials on financing with water rights, sandstone rights, forest rights, carbon sink rights and cultural IP and other rights and interests of ecological resource assets as collaterals, and be responsible for innovation of green finance products with the help of financial institutions. The "Liangshan Bank" reform in Quzhou City puts focus on the "pain points" and "difficult points" including high transaction costs, difficulty in introducing industries, numerous development barriers, and difficulty in financing and ecological realization in the process of transforming ecological resources into assets and capital. Quzhou City has established a credit enhancement system through creating the "Liangshan Bank" financial service, and empowered the operation right and means of production of the main body of ecological resources and other right items subject to paid acquisition through finance. It has effectively used and realized precipitated resources, propelled ecological resources to combine with financial products, and pushed forward the transformation of ecological resources into ecological assets through guarantee, promised acquisition, prioritized disposal and other forms, providing a path for "lucid waters and lush mountains" to convert to "invaluable assets".

Demonstration section of "Shilichangfeng", a picturesque scenic zone in Changshan county *Photo Credit* Bureau of Agriculture and Rural Affairs of Quzhou City

4.9.3 Key Achievements

4.9.3.1 Reducing Carbon Emissions from the Production End to the Consumption End

Up to now, three major carbon account systems have covered a total of over 600 industrial enterprises, more than 180 large planters and farmers and more than 1.37 million personal accounts. The construction of industrial enterprise carbon accounts has covered more than 600 enterprises in traditional industry sectors such as chemical, machinery, building material, papermaking, smelting and pressing of ferrous metals, and energy with a proportion of over 70% in the industrial output value of industrial enterprises above designated size and a proportion of 90% in carbon emissions in Quzhou City. The first batch of construction of agricultural carbon neutrality accounts has covered 10.7% of pig farms, 20% of major crop planters and all organic fertilizer production enterprises. Since the launching of personal carbon accounts, 5,290,000 personal carbon accounts have been allocated, and a total of 5230 tons of carbon emissions have been reduced.

4.9.3.2 Making Remarkable Achievements in the Transformation of Traditional Industry Sectors

In 2019, Quzhou City introduced 10 "Zhejiang Made" standards, 13 enterprises with "Defined Quality", and 9 provincial-level digital transformation pilot demonstration enterprises. It newly cultivated 3 "hidden champion" enterprises and 1 "single product champion" enterprise, closed down 32 enterprises with insufficient production capacity and overcapacity, and rectified 138 enterprises (workshops) working with a low industrial level, scattered layout and disordered management. It further eliminated 70 enterprises with backward production capacity and rectified 431 enterprises with a low industrial level, scattered layout and disordered management, and shut down 31 "zombie enterprises", effectively reclaiming industrial land of 5,280 mu (or 352 ha). The tax revenue per mu of industrial enterprises above the designated size increased by 23.3%, while investment in the manufacturing industry of Quzhou City grew by 15.2%. In the first half of 2020, four products from three enterprises in Quzhou City passed the "Defined Quality" brand certification and the proportion of sales from branded enterprises and enterprises implementing "Zhejiang Made" standards reached 31.3%, representing an increase of about 4% compared with the previous year. As of the end of July 2020, Quzhou City had certified 531 agricultural products as "pollution-free agricultural products, green food, organic agricultural products, or agricultural products with geographical indications", including 169 organic agricultural products, 75 green foods, 282 pollution-free agricultural products, and 5 agricultural products with geographical indications. As of the end of March 2021, the citywide green credit balance across Quzhou City was CNY118.158 billion, nearly 5 times that of the end of 2016 prior to green finance reform. This figure also accounts for 36.46% of all loan balances, strongly supporting transformation and upgrading as well as green development of local traditional enterprises.

Modern agricultural garden of the Longkou tea industry *Photo Credit* Bureau of Agriculture and Rural Affairs of Quzhou City

4.9.3.3 Activating Rural Ecological Resources to Increase the Income of and Enrich Farmers

"Liangshan Bank" in Changshan County opened for operation formally in September 2020. It has implemented 30 ecological transformation cases, reciprocated 176 village collectives, collected and stored investments of CNY350 million, extended credits of CNY134 million to 213 borrowers, and issued ecological loans of CNY132 million. With carbon sink value generated from 5,244 mu (or 349.6 ha) of forest resources as its pledge, Huayuan Village, Majin Town, Kaihua County has secured a pledged loan of CNY3 million from the Agricultural Bank of China in Kaihua County as start-up capital for construction of a camellia oleifera base renovation project. "Liangshan Bank" in Kecheng District has helped solve planting and sales problems after collecting and storing nearly 1000 mu (or 60 ha) of waste land in Yuanlin Village of Huashu Township, with income shared fifty-fifty. It is expected that the village collective and villagers will see an annual income increase of CNY600,000. Abundant ecological resources in Quzhou City are becoming increasingly valuable under the push of "Liangshan Bank", and new development ideas and opportunities are being generated therefrom.

Genyuan town in Kaihua *Photo Credit* Zhejiang Provincial Department of Culture and Tourism

4.9.4 Inspirations

Located at the source of the Qiantang River, in order to add new drivers to economic development and protect the ecological barrier of Zhejiang Province, Quzhou must pursue green development as well as transformation and upgrading. Therefore, focusing clearly on "supporting green transformation and transition of traditional industries with finance", Quzhou City has proposed to create a "pollution-free and clean" new development mode of ecological industry. Persisting in the "green and feature" principle, it has guided financial institutions to organically integrate green development with business management and innovate service modes, striving to build a Quzhou mode featuring "big garden leadership, big platform gathering, big data support and extensive service linkage". The "green finance reform" mode of Quzhou City applies to traditional heavy chemical industrial bases with important ecological locations, rich ecological resources, as well as heavy chemical industries and hazardous chemical product production enterprises. This mode facilitates the green transition of traditional industries and unleashes their new vitality through green finance, which in return reciprocates the ecology. It effectively uses ecological resources and pushes forward their value transformation, blazing a new green development path.

4.10 Huzhou City (Zhejiang Province)—Introducing Diversified Green Finance Services to Promote High-Quality Development

4.10.1 Background

Located in the central area of the Yangtze River Delta, Huzhou City enjoys convenient transportation link and clear regional advantages. It is the shared hinterland of Shanghai, Hangzhou, and Nanjing, and also a node city connecting the north and south wings of the Yangtze River Delta and the eastern and central regions of the delta. Huzhou has jurisdiction over two districts and three counties, covering an area of 5820 km^2, with a registered population of 2.65 million and a permanent population of 2.98 million. It registered a GDP of CNY312.24 billion in 2019. Huzhou City, which is home to a whole host of historical figures and enjoys an abundance of cultural heritage, is hailed as the "production base of silk", "land of fish and rice", and "cultural center". It is an attractive city with beautiful mountains and rivers and a charming ecological environment. Huzhou boasts itself as the birthplace of the concept of "Lucid Waters and Lush Mountains are Invaluable Assets" and the birthplace of "beautiful countryside in China". The city is also the pioneer of "ecology+" practice in China and an ecological conservation land of the Taihu Lake Basin.

In recent years, in an effort to meet the requirements of the goal of creating a "Huzhou sample and Huzhou model", Huzhou has been proactively exploring and carrying out a series of innovations and practices with local characteristics. This includes integrating green finance with green manufacturing, environmental governance, inclusive finance, rural revitalization, social governance, and financial technology. Huzhou City has also won a number of honorary titles, including "National Environmental Protection Model City"; "National Sanitary City"; "National Garden City"; "Top Tourism City of China"; "China Charm City"; "National Top 100 Cities in Comprehensive Strengths"; "National Forest City"; and "China's Happiest City". In addition, Huzhou has become China's first prefecture-level zone that demonstrates best practices for promoting a sound ecology.

4.10.2 Specific Practice

4.10.2.1 Green Finance + green Smart Manufacturing to Activate the Financial Engine of Green Industry

In a bid to precisely connect green finance with green smart manufacturing standards, Huzhou City has incorporated "345 + 1.5" (i.e., the average tax revenue per mu, or about 667 m^2, is CNY300,000, the average fixed asset investment intensity per mu is CNY4 million, the average output value per mu is CNY5 million, and the building plot ratio is no less than 1.5), benefit per mu, and other requirements into green finance standards and created a system covering 12 local standards for green finance. This system, which can precisely identify the "green" and "non-green" nature and green grades of borrowers, is linked with policies and measures such as financial discounts and credit interest rates to guide financial resources to lean towards green smart manufacturing fields. In terms of financial discounts, depending on the grades of "dark green", "medium green", and "light green", green enterprises (projects) are granted a loan with discounted interest with a benchmark interest rate of 12%, 9% and 6%, respectively. In terms of bank credit, Huzhou has created the "six separations" (separation of credit scale, approval channels, fund price, risk management indicators, performance assessment, and of green financial products) franchise model for green credit. For example, Nanxun Bank has set up the first green branch in China, implementing a 10% interest rate discount for green loans of the same grade under the same category. Huzhou Branch of Agricultural Bank of China has also developed the "series of loans for green enterprises (projects)", which grant preferential loans that may be as low as benchmark interest rate to green enterprises. This series of green loans also provide a certain proportion of credit loans.

Customized financial services supports Anji white tea *Photo Credit* Huzhou Municipal People's Government

4.10.2.2 Green Finance + Environmental Governance to Build an Ecological Protection Supervisory System

Huzhou City pushes financial institutions to fully implement the "one-vote veto system for environmental protection projects", "negative list on environmental protection", and other credit management mechanisms, organically combining the allocation of resources with "energy consumption" and "pollution. Loans to industries with high pollution, high energy consumption and overcapacity have dropped over the past three consecutive years. Huzhou has established a multi-level environmental information disclosure mechanism and issued China's first three-year plan on environmental information disclosure for financial institutions, in an effort to guide financial institutions to effectively fulfill their social responsibilities for environmental protection. The city has implemented a financial support plan for volatile organic compounds (VOCs) emission reduction, and in 2020 it arranged CNY1 billion in preferential loans to support enterprises' VOCs emission reduction efforts in a targeted manner. Huzhou also makes efforts to leverage the role of insurance in environmental governance, supervision and risk management. It pioneered in China the environmental pollution liability insurance model featuring "insurance, service, supervision and credit", under which 50% of the premium is used for environmental risk assessment and the assessment results are linked to environmental supervision, underwriting rates, and credit access, so as to push enterprises to improve their environmental risk management capabilities.

Abandoned mines becoming scenic spots *Photo Credit* Huzhou Municipal People's Government

4.10.2.3 Green Finance + Inclusive Finance to Strike a Balance Between Stable Growth and Ecological Protection

Traditional small and micro enterprises are an indispensable part of maintaining supply chain security. In order to solve the problem of green development facing small and micro enterprises, Huzhou has introduced an innovative model of "industrial park loans", established green small and micro parks, and guided small and micro enterprises to cluster in the parks, providing unified environmental protection and energy infrastructure. The bank provides a full set of financial services such as development loans, fixed asset purchase loans, and operational current loans for the ex-ante park construction, the interim entry of enterprises into the park, and the ex-post production and operation of enterprises. Meanwhile, Huzhou has established guarantee systems of "green loan insurance" and "green loan guarantees", innovated a bottom-line credit loan risk compensation mechanism, and launched more than 30 loan turnover products of "renewal loans", in a bid to push banks to "dare to lend" and "extend loans fast" to small and micro enterprises.

Construction waste becoming "new environmentally-friendly bricks" *Photo Credit* Huzhou Municipal People's Government

4.10.2.4 Green Finance + Rural Revitalization to Implement Financial Drip Irrigation Projects, Integrating Green Finance with Beautiful Rural Construction

In Zhejiang Province, Huzhou took the lead in launching innovative green insurance products such as "green agricultural product price index insurance" and "distinctive agricultural product temperature index insurance", in an effort to use green insurance tools to compensate farmers' losses, prevent poverty due to disasters, and serve rural revitalization. It also carried out green asset securitization relying on the balance sheet of natural resources, and promoted innovations in pledge and collateral business including future usufruct and rural "two rights" (the right to operate contracted land in rural areas and the property rights of farmers in housing). With a view to improving its green finance guidance mechanism, Huzhou City has closely combined the reform and innovation of green finance with major strategies such as facilitating rural revitalization and the construction of the pilot zone of "Lucid Waters and Lush

Mountains are Invaluable Assets". A database covering over 200 credit products in key areas of financial support has been established, and a series of new financial support products such as "photovoltaic loans", "agricultural and forestry loans", and "homestay loans" have been innovatively developed. The guidance and guarantees of green finance have promoted various national pilot construction projects and implementation of major strategies.

The "Red Code" of Xiantan Village *Photo Credit* Huzhou Municipal People's Government

4.10.2.5 Green Finance + Social Governance to Form a Working Method "in Advance of Risk"

Huzhou City took the lead in Zhejiang Province to establish a set of "5 + X" financial linkage mechanisms featuring "coordination among departments, smooth operation, and each performing its own duties", and took the initiative to fight and resolutely win the battle against internet financial risks. Huzhou also proactively issued policy opinions on "nine bailout measures" for listed companies, and pioneered the seven bailout models of "financial incentives, market debt, stock and debt portfolios, financial support, specialized service teams, active justice, and expanded financing". The city has established a classified assistance mechanism for distressed enterprises, compiled the "three lists", and provided precise and classified assistance, so as to effectively maintain mutual trust between banks and enterprises and to stabilize enterprise operations. Finally, through the "financial mediation and judicial confirmation" mechanism, the Green Finance Dispute Mediation Center enables green finance dispute cases to be resolved efficiently and at low cost, achieving a win–win situation for the four parties of "government, banking, enterprises and court".

4.10.2.6 Green Finance + Fintech to Create a "One-Stop" Finance Integrated Service Platform

Huzhou has established the Huzhou green finance integrated service platform, providing enterprises with "one-stop" financial services such as loans, equity, guarantees, policy fulfillment, and green identification. This platform helps enterprises to "online shop" loans from 36 banks in the same way that one would shop on Taobao, to have their online roadshows recommended to more than 100 investment institutions and investors across China, and to apply for guarantees to all policy guarantee institutions in Huzhou City with one click. The platform has pioneered a "three services (order-grabbing service, limited-time service, and door-to-door service)" mechanism. By integrating and sharing government data of over 1000 fields from 34 units, the platform allows fewer errands for enterprises and fast approval for banks, thereby substantially shortening the time enterprises spend waiting for their loans. In addition, by innovatively using fintech, Huzhou has also set up China's first green finance information management system and green bank evaluation system.

4.10.3 Key Achievements

4.10.3.1 Green Development Index Ranks Among the Top

With the support of green finance, by the end of 2020, Huzhou had cumulatively provided CNY18.7059 million in discount interest for green enterprises (projects). Meanwhile, environmental pollution liability insurance has achieved 100% coverage in 8 key industries and 316 high-risk environmental enterprises in the city. Through environmental risk assessments, a total of 2590 hidden environmental risks have been detected and addressed. Since small and micro enterprises entered the park, on an annual basis, sewage discharge has been reduced by 1 million tons, electricity by 13 million kW·h, and smoke emission by 126,000 tons. The green development index of Huzhou City has ranked second in Zhejiang Province for two consecutive years, and the growth rate of industrial added value of the three leading industries—high-tech industry, strategic emerging industry, and equipment manufacturing industry—all rank first in Zhejiang Province.

4.10.3.2 Environmental Resources Achieve "Financialization"

Huzhou has developed 180 + innovative green finance products. Anji County first introduced the model of "Lucid Waters and Lush Mountains are Invaluable Assets—Agricultural and forestry loans" featuring win–win cooperation between government, banks, and farmers, in a bid to enliven agricultural production resources with green credit and boost rural revitalization. Under this model, a total of CNY1.582 billion in financial services for agriculture-related entities has been provided, and this model has been gradually promoted in the agricultural credit system in Zhejiang Province. As of the end of May 2020, loans to small and micro enterprises in Huzhou had increased by 25.29% year-on-year, 6.48 percentage points higher than the growth of Zhejiang Province. The balance of agriculture-related loans of bank institutions in Huzhou was CNY302.929 billion, an increase of 23.05% year-on-year, accounting for 54.94% of all loans.

4.10.3.3 A Variety of Means Are Used to Build a Security System

Over the two years since the launch of the green finance integrated service platform, it has helped more than 16,000 enterprises to obtain CNY179.262 billion in bank credit, with a total of CNY7.08 billion in financing granted to 75 science and technology innovation projects. The platform has effectively made access to financing easier, cheaper, and faster for enterprises. As of the end of May 2020, the non-performing loan rate of banks in Huzhou was 0.5%, the city's credit asset quality ranked second in Zhejiang Province, and the green credit non-performing rate of the city remained at a low level of below 0.04%. Powered by its green finance information management system and green bank evaluation system, Huzhou has for the first time achieved traceability, comparison, and measurability of green credit data and T + 0 number counting, as well as online and smart evaluation on the performance of banks in green finance.

4.10.4 Inspirations

While developing green finance, Huzhou consistently works towards the goal of "green finance+". To this end, it first establishes systems and standards, creates a good development environment, then builds a platform for bank-enterprise connection and strengthens competition among banks.

According to standards, Huzhou classifies financing enterprises and projects into different green grades to provide reference for banks to implement green financial preferential measures such as discount interest subsidies. Huzhou also strengthens government-bank cooperation, expands the innovation of green financial products, launches a wide variety of green financial innovation products in a timely manner, and provides support for the government to promote green development transformation. This development model of Huzhou, which features embracing innovation to foster "green drivers", is suitable for areas with an excellent ecological environment and developed economy. On the premise of achieving sufficient results in economic construction, a city can innovate and develop green finance through financial means to revitalize ecological resources and achieve sustainable development of green ecology, economy and society.

Chapter 5
Evaluation with LLEI

Carrying out evaluation and research on "Lucid Waters and Lush Mountains are Invaluable Assets" ("Liangshan" for short) is an important measure and practical need for promoting the nationwide practice of "Lucid Waters and Lush Mountains are Invaluable Assets", strengthening the construction and management of "Lucid Waters and Lush Mountains are Invaluable Assets" practice and innovation bases, and for guiding, diagnosing, and calibrating local "Liangshan" practice and green development. At present, establishing a set of scientific and reasonable evaluation methods and technical standards is urgently needed, covering how to scientifically and quantitatively measure the effectiveness of "Liangshan" construction, analyze its process and level, and to guide and promote green development. In light of the technical support work for "Lucid Waters and Lush Mountains are Invaluable Assets" practice and innovation bases and on the basis of the preparation of the *Management Regulations for the Construction of "Lucid Waters and Lush Mountains are Invaluable Assets" Practice and Innovation Bases (for Trial Implementation)* and the evaluation's technical guidelines, we have proposed the evaluation framework and index system of "Lucid Waters and Lush Mountains are Invaluable Assets" Evaluate Index, LLEI", in order to evaluate the practice effectiveness of "Lucid Waters and Lush Mountains are Invaluable Assets".

5.1 LLEI Construction

The "Lucid Waters and Lush Mountains are Invaluable Assets" Evaluate Index (LLEI) is a comprehensive index that reflects regional ecological environmental quality and ecological asset status, the transformation degree of lucid waters and lush mountains into invaluable assets, the guarantee degree of "Liangshan" and the transformation path of "Liangshan". LLEI also serves to quantify the construction level

H. Zhang et al., *Theory and Practice of "Lucid Waters and Lush Mountains Are Invaluable Assets" in China*, https://doi.org/10.1007/978-981-97-7340-4_5

and effectiveness of "Liangshan" to guide and promote green development (Zhang Huiyuan 2020).

5.1.1 Inherent Logic

Guided by the concept of "Lucid Waters and Lush Mountains are Invaluable Assets", the LLEI should reflect the evolving law of the relationship between economic and social development and ecological environmental protection, and systematically integrate the four aspects of ecological environment, economy, society and people's wellbeing. Furthermore, the LLEI should clarify that lucid waters and lush mountains are not only natural wealth and ecological wealth, but also social wealth and economic wealth, and should reveal the truth that to protect the environment is to protect productivity, and that to improve the environment is to develop productivity.

The LLEI should also quantitatively express the effectiveness of "Liangshan" construction. Not only should it intuitively quantify the transformation degree, guarantee degree and construction level and process of "Liangshan" construction, but it should also consider the goal of "Lucid Waters and Lush Mountains are Invaluable Assets" construction, and continue to promote the work by setting specific indicators.

Finally, the LLEI should help guide the development direction of "Liangshan" construction and explore the transformation path of "Liangshan". It should reflect the direction of guiding green development, striving to promote the development of "ecology and industry", and promoting ecological industrialization and industrial ecologicalization. Meanwhile, the LLEI should reflect the realization method and transformation path of the value of ecological products, striving to transform the ecological environment advantage of lucid waters and lush mountains into the economic development advantage brought by invaluable assets.

5.1.2 Indicator Selection

Based on the concept connotation of the LLEI and focusing on the goal, direction and task of "Liangshan" construction, we have established an LLEI evaluation indicator system using a top-down and layer-layer breakdown method. The system, which reflects all-factor ecological value, multi-path transformation channels, and equal support from the government and market, can be divided into the "Lucid Waters and Lush Mountains" Index, "Liangshan Transformation" Index and "Liangshan Guarantee" Index. There are 20 specific indicators in 8 tasks, including environmental quality, ecological regime, people's wellbeing, ecological economy, ecological compensation, social benefits, institutional innovation, and financial security (Fig. 5.1).

The "Lucid Waters and Lush Mountains" index focuses on the consideration of regional ecological environmental advantages and ecological assets, corresponding

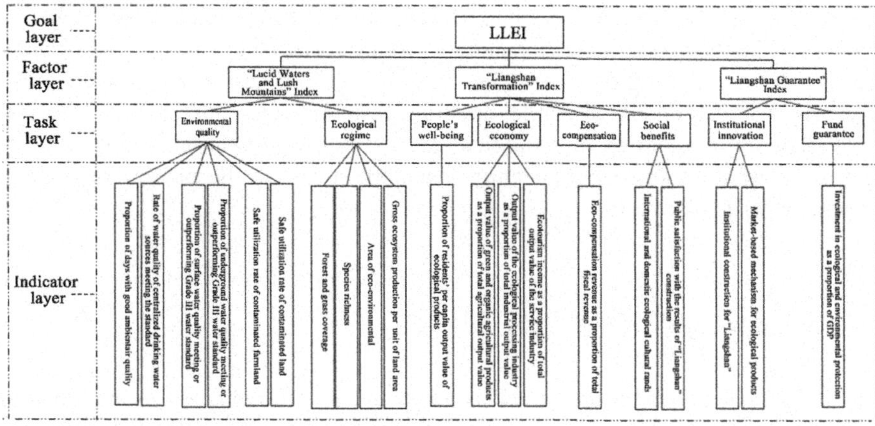

Fig. 5.1 LLEI evaluation index framework system

to protecting lucid waters and lush mountains, improving the supply of ecological products and maintaining and increasing the value of natural capital. The value of "Lucid waters and Lush Mountains" is associated closely with the quantity, quality, and structure of the relevant elements of the ecological environment, so indicators in ecological regime and environmental quality are mainly considered when selecting indicators. The selection of environmental quality indicators should take into account all elements regarding water, air and soil, while key indicators representing environmental quality should be selected with quality indicators as the core. Ecological regime indicators should be selected to reflect the status of the ecosystem, the degree and scale of ecological protection, as well as the indicators showing the supporting role of the ecosystem in improving people's quality of life.

The "Liangshan Transformation" Index focuses on considering the path, benefits and degree of transformation from "Lucid Waters and Lush Mountains" into "Invaluable Assets", and the corresponding improvements in green development level and the wellbeing of various groups involved. It primarily includes four aspects: people's wellbeing, ecological economy, ecological compensation, and social benefits. In the selection of indicators, we should consider selecting indicators that can reflect the transformation path of "Liangshan", including ecological agriculture, ecological industry, ecological service industry, and ecological compensation. However, in consideration of people's wellbeing and social benefits, we should also select indicators reflecting the contribution of "Liangshan" transformation to the wellbeing of public individuals and society as a whole.

The "Liangshan Guarantee" Index focuses on the degree of guarantee of institutional innovation and capital investment for the protection of "Lucid Waters and Lush Mountains" and the "transformation of Liangshan". It involves two key aspects: institutional and mechanism innovation and capital investment. Of the two, institutional and mechanism innovation takes into account market-based "Liangshan" transformation mechanisms such as carbon emission rights, pollutant discharge rights, water rights, forest rights, and green finance, as well as the value realization mechanism of ecological products. Meanwhile, capital investment represents the local government's degree of attention to ecological environmental protection, and reflects the repayment of the transformation into invaluable assets to the protection of lucid waters and lush mountains.

5.2 Evaluation Method for LLEI

The LLEI is calculated by a weighted average method, with a higher evaluation score indicating a better construction level and a more prominent construction effectiveness. In order to eliminate the impact of different index units on the evaluation results, an extreme value standardization method is used to nondimensionalize the indicators. The extreme value should consider not only regional differences, but also the requirements of construction objectives. It can be determined in light of the characteristics of the original green ecology, resource and environmental carrying capacity, social and economic development, and published policy documents. In addition, in order to better reflect the relative importance of relationship between indicators, the weight of indicators is determined by analytic hierarchy process. The LLEI is calculated with the following method:

$$\begin{aligned} \text{LLEI} &= \sum_{i=1}^{3} \omega_i \cdot Y_i \\ &= \omega_1 \times \sum_{i=1}^{2}\sum_{j=1}^{m} \alpha_{ij} \cdot G_{ij} + \omega_2 \times \sum_{i=1}^{4}\sum_{j=1}^{m} \beta_{ij} \cdot T_{ij} + \omega_3 \times \sum_{i=1}^{3}\sum_{j=1}^{k} \gamma_{ij} \cdot M_{ij} \end{aligned}$$

In the above formula, ω_i is the weight of the "Lucid Waters and Lush Mountains" Index, "Liangshan Transformation" Index, and "Liangshan Guarantee" Index; α_i is the weight of the two tasks of environmental quality and ecological regime; α_{ij} is the weight of the indicators under these two tasks; G_i is the value of the two tasks of environmental quality and ecological regime; G_{ij} is the relative value of the corresponding indicators; β_i is the weight of the four tasks of people's wellbeing, ecological economy, ecological compensation, and social benefits; β_{ij} is the weight of the indicators under these four tasks; T_i is the value of the four tasks; T_{ij} is the value of the indicators under the four tasks; γ_i is the weight of the two tasks of institutional innovation and financial security; γ_{ij} is the weight of the indicators under the two tasks; M_i is the value of the two tasks, and M_{ij} is the value of the indicators under the two tasks.

5.3 Application of LLEI

The LLEI serves as a comprehensive analysis tool for evaluating, diagnosing, calibrating, and guiding the practice of "Lucid Waters and Lush Mountains are Invaluable Assets" and as an important reference for the post-evaluation and management of "Lucid Waters and Lush Mountains are Invaluable Assets" practice and innovation bases. It is mainly used to guide local governments to clarify construction objectives, directions, key tasks, summarize and condense transformation path models, quantify construction levels, dynamically calibrate the direction of construction, and innovate and explore the institutional practice and action practice of transformation. The LLEI is primarily reflected in the following three aspects.

The LLEI can help to horizontally compare and analyze construction effectiveness. According to the evaluation results of LLEI and the three sub-indexes, we can compare and analyze the horizontal differences, convergence and effectiveness of different localities to objectively reflect construction effectiveness. Furthermore, based on the evaluation results, we can conduct classified research according to the principle of similarity, to examine levels of construction, transformation paths, and experience models. At the same time, we can select places with high scores as typical cases and model cases to further condense the path models of transformation, summarize and analyze applicability, transformation conditions, and inspirations.

The LLEI can also serve to longitudinally diagnose construction processes. Local governments can carry out self-assessments of construction effectiveness, comprehensively analyze the annual scores of the LLEI, the "Lucid Waters and Lush Mountains" Index, "Liangshan Transformation" Index and "Liangshan Guarantee" Index, and longitudinally analyze the inter-annual changes. They should promptly discover problems and shortcomings in the practice process, deeply analyze the change mechanisms and development trends, and use the evaluation results as an important basis for dynamic optimization and adjustment.

Finally, the LLEI can be used towards scientifically guiding local practice and application. In the process of carrying out practical exploration of "Lucid Waters and Lush Mountains are Invaluable Assets", promoting the construction of "Lucid Waters and Lush Mountains are Invaluable Assets" practice and innovation bases, and formulating specific implementation plans, action plans, and annual work plans, local governments can refer to the LLEI to set characteristic target indicators and clarify tasks in terms of protecting lucid waters and lush mountains, transforming into invaluable assets, ensuring long-term mechanisms, and sharing green benefits with local people. The LLEI can guide local governments to strike a balance between protection and development, protect the potential and sustainability of economic and social development, and push for the formation of a green development model, and explore a green and high-quality development path.

5.4 Empirical Analysis of LLEI

5.4.1 Saihanba Mechanized Forest Farm

The Saihanba Mechanized Forest Farm is among the first batch of "Lucid Waters and Lush Mountains are Invaluable Assets" practice and innovation bases named in 2017, and it is also the first such practice and innovation base with a forest farm as its concept, and therefore has significant representative and regional characteristics.

According to the calculation, the LLEI of Saihanba Mechanized Forest Farm from 2017 to 2019 (Table 5.1 and Fig. 5.2) was 79.78, 82.70, and 85.62, respectively. Specifically, from 2017 to 2019, the forest farm's "Lucid Waters and Lush Mountains" Index remained stable at 87.20; its "Invaluable Assets" Index was 75.60, 78.80, and 82.00, respectively; and its "Long-term Guarantee" Index was 76.00, 82.00, and 88.00, respectively.

The development characteristics of "Liangshan" in this area are as follows.

Table 5.1 LLEI evaluation results of Saihanba mechanized forest farm

Index	2017	2018	2019
Comprehensive index	79.78	82.70	85.62
Lucid waters and lush mountains index	87.20	87.20	87.20
Invaluable assets index	75.60	78.80	82.00
Long-term guarantee index	76.00	82.00	88.00

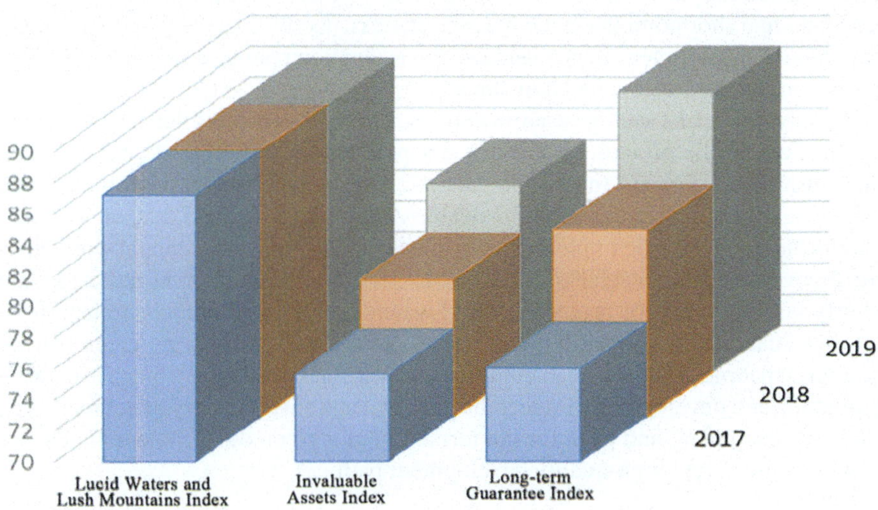

Fig. 5.2 Evaluation results of sub-indexes of Saihanba mechanized forest farm

5.4.1.1 A Stable Natural Ecosystem, with Ecological Regime and Environmental Quality Reaching an Excellent Level

Saihanba Forest Farm is blessed with a good ecological environment and rich forest resources. In 2020, the proportion of days with good ambient air quality reached 100%; the coverage rate of forest and grass reached 82%; the gross ecosystem production per unit of land area was CNY23 million/km^2; and the negative ion content in the air was six times that of the average for large and medium-sized cities. Annually absorbing 14,600 tons of sulfur dioxide and nitrogen oxides every year, and trapping 156,400 tons of dust, the forest farm serves as an important ecological barrier for the Beijing-Tianjin-Hebei region. This excellent ecological environment has laid a good foundation for local practice and exploration of "Liangshan".

5.4.1.2 The Ecological Industry Has a High Level and Great Potential for Development

Saihanba Forest Farm has abundant forest resources, vast grasslands and unique cold, snow and ice resources, as well as wetland resources. Ecological tourism has become a dominant local industry, and green seedlings and forestry carbon sinks are booming. Employees on forest farms earn an average annual income of nearly CNY100,000, driving the development of green industries such as farmhouse tours, rural tours, handicrafts, the picking industry, native product processing, and aquaculture. A total social income of more than CNY600 million is achieved every year. Therefore, the "Liangshan" transformation has achieved great economic and social benefits.

5.4.1.3 Prominent Brand Advantage of Ecological Culture, Playing a Significant Role in Promoting the Development of the Ecological Industry

Saihanba Forest Farm holds a large number of international and domestic ecological cultural brands. Since 2017, it has successively won the highest honor of the United Nations on environmental protection—"Champion of the Earth Award"; the title of "The Most Beautiful Fighter" awarded by nine ministries and commissions such as the Central Publicity Department and the Central Organization Department; the title of "2018–2019 Green China Person of the Year" by six departments such as the Environmental Protection Commission of the National People's Congress and the Ministry of Ecology and Environment; and the title of "National Poverty Alleviation Model". As Saihanba grows significantly in visibility and influence, its entrepreneurial enthusiasm has been fully stimulated. The endogenous development driver of the ecological industry, coupled with external environmental advantages, has injected more momentum into the development of Saihanba's ecological industry.

5.4.1.4 Greatly Improving Long-Term Institutional Guarantees, Leading to Further Exploration in Market-Based Mechanisms for Ecological Products

In 2017, Saihanba Forest Farm's afforestation and forestry carbon sink project became the first forestry carbon sink project of a state-owned forest farm successfully registered and issued by the National Development and Reform Commission. In 2018, Saihanba Forest Farm concluded the first transaction of 36,000 tons of forestry carbon sinks with Beijing Lannuo Century Technology Co., Ltd. in Beijing Environmental Exchange, achieving an income of CNY900,000. In 2019, the forest farm traded 2981 tons of carbon sinks with Shell Energy Co., Ltd., generating an income of CNY39,000. In the new development stage, forestry carbon sinks, as a new type of business spawned by the "carbon peaking and carbon neutrality (dual carbon)" goals, will enjoy a successful trading form and development trend going forward. According to accounting, Saihanba's forest assets have a total value of CNY20.6 billion and annually fix 814,100 tons of carbon dioxide. Thanks to the "dual carbon" strategy, Saihanba's ecological products face unprecedented historical opportunities and many policy dividends for the exploration of a market-based mechanism, and therefore enjoy huge development potential.

5.4.2 Huzhou City

Huzhou boasts itself as the birthplace of the concept of "Lucid Waters and Lush Mountains are Invaluable Assets" and the birthplace of "beautiful countryside in China". It is also an ecological conservation land of the Taihu Lake Basin. Huzhou has achieved remarkable results both in promoting the construction of ecological civilization and practicing the concept of "Lucid Waters and Lush Mountains are Invaluable Assets", having carved out a path of socialist ecological civilization construction with Chinese characteristics featuring beautiful ecology, thriving industries, and well-to-do local people.

According to calculations, the LLEI of Huzhou City from 2017 to 2019 (Table 5.2 and Fig. 5.3) was 73.18, 79.77, and 85.61, respectively. Specifically, from 2017 to 2019, the forest farm's "Lucid Waters and Lush Mountains" Index remained stable at 85.60; its "Invaluable Assets" Index was 68.62, 72.03, and 82.00, respectively; and its "Long-term Guarantee" Index was 64.00, 82.00, and 88.00, respectively.

The development characteristics of "Liangshan" in this area are as follows.

Table 5.2 LLEI evaluation results of Huzhou city

Index	2017	2018	2019
Comprehensive index	73.18	79.77	85.61
Lucid waters and lush mountains index	85.60	85.60	85.60
Invaluable assets index	68.62	72.03	82.00
Long-term guarantee index	64.00	82.00	88.00

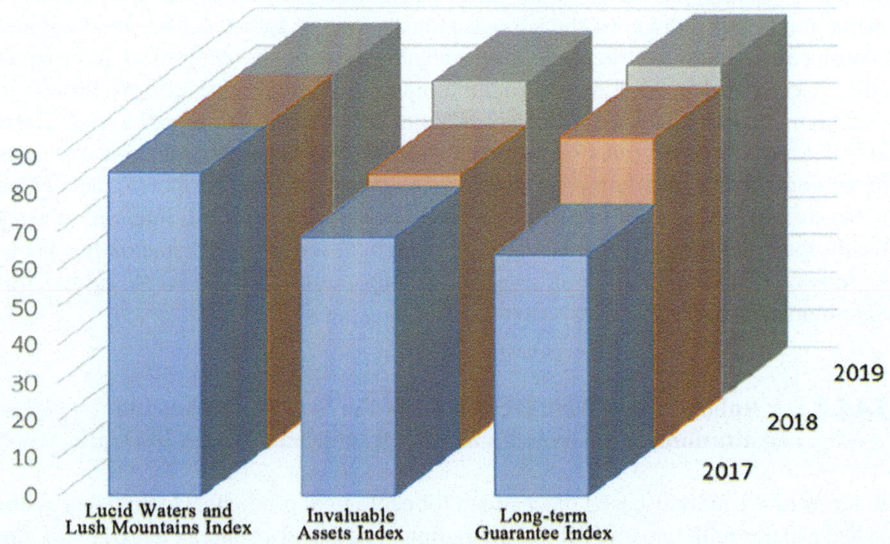

Fig. 5.3 Evaluation results of sub-indexes of Huzhou city

5.4.2.1 Prominent Advantages of Ecological Resources, and a Steadily Improving Ecological Regime and Environmental Quality

Huzhou City is located on the Hangjiahu Plain, and more than 80% of the arable land in the city is high-yielding farmland that ensures stable yields despite droughts or excessive rain. Huzhou is a main production area and important production base of grain, cocoons, freshwater fish and moso bamboo not only for Zhejiang Province, but for China as a whole. Linghu District is one of the three major freshwater fish breeding bases in China, while Anji County ranks first among China's top ten "towns of moso bamboo", with unique advantages in ecological resources. Since 2017, the proportion of surface water quality meeting or outperforming Grade III water standard has remained at 100%, and the proportion of days with good ambient air quality has increased from 68.5% to 76.7%. The requirements of heavy metal pollution reduction targets have been met ahead of schedule, and the target of zero growth in fertilizer use has been completed six years ahead of schedule.

5.4.2.2 The Ecological Industry is Booming in Certain Sectors, and the Overall Level of "Transformation into Invaluable Assets" Still Has a Large Room for Improvement

In Huzhou City, the output value of green and organic agricultural products accounts for more than 60% of the total agricultural output value, but the per capita output value of ecological products, the output value of the ecological processing industry as a proportion of total industrial output value, and the ecotourism income as a proportion of the total output value of the service industry are all below 20%. The dominant ecological agriculture reflects that the development of the ecological industry is still concentrated in rural areas. Yucun Village in Anji County, Longxi Village in Changxing County, Digang Village in Nanxun Town, and other areas have achieved excellent results in the transformation of "Liangshan". However, Huzhou still faces an ecological industry defined by unbalanced urban–rural development, as well as by an overall low level of industrial chain extension and industrial integration. As a result, the potential of the advantages of ecological resources in Huzhou has yet to be fully realized, and the overall level of "transformation into invaluable assets" still has significant room for improvement.

5.4.2.3 A Robust Long-Term Guarantee of the System, Advancing Institutional Construction for the "Liangshan" Bases in Depth

In terms of the market-based mechanism for ecological products, Huzhou has established a major pollutant emission right trading system and an energy use right trading system, as part of its effort to actively explore the reform and innovation of green finance. On institutional construction for the bases, Huzhou has established a "three-in-one" institutional guarantee system of legislation, standards and mechanisms, using the "two standards and one certification" mechanism to promote the green development of industries and to revitalize rural institutions and mechanisms. The Long-term Guarantee Index has achieved significant improvements on the basis of good levels. Institutional construction for the "Liangshan" bases has seen no decline in terms of both in-depth advancement and innovative exploration, proving its advanced and demonstrative natures.

Chapter 6
Prospect

Different localities in China are actively carrying out institutional innovation and practical exploration of "Lucid Waters and Lush Mountains are Invaluable Assets". These areas have made significant progress and remarkable results in ecological protection and restoration and supply of ecological products, the transformation paths of "Lucid Waters and Lush Mountains are Invaluable Assets", economic transformation and upgrading and high-quality development, and the establishment and improvement of the value realization mechanism of ecological products. In the process of practice, the transformation paths of "Lucid Waters and Lush Mountains are Invaluable Assets" in various localities feature both cross-combination and evolution and substitution, which is testament to the internal unity, mutual promotion and coordinated symbiosis between ecological environmental protection and economic development. In order to further clarify the transformation path mechanism of "Liangshan", evaluate the transformation effect of "Liangshan", form demonstration models of "Liangshan" transformation, and to leverage the leading role of models, it is necessary for us to strengthen the following four aspects of research.

We should strengthen theoretical research and encourage science and technology to "empower green". Based on management and service decision support, we should strengthen research on the basic theory, evaluation framework system, evaluation methods and indicators regarding ecological product value realization. We should also consider formulating ecological product value accounting standards or norms, and form standardized and unified ecological product value accounting methods and standards. In response to problems such as unclear transformation paths, imperfect transformation mechanisms, unclear optimal path modes, and difficulties in quantifying the value of ecological products, we should set up a special research plan to tackle these issues. Local governments, enterprises and social forces should be encouraged to invest in the theoretical and practical research of "Liangshan" and to build a number of industry-university-research bases for the practice of "Liangshan".

We should also strengthen effectiveness evaluation to guide green development. We should research and formulate evaluation technical specifications, research and

H. Zhang et al., *Theory and Practice of "Lucid Waters and Lush Mountains Are Invaluable Assets" in China*, https://doi.org/10.1007/978-981-97-7340-4_6

propose a systematic and scientific LLEI evaluation system in terms of specific processes, methods and applications, and further clarify normative reference documents, definitions of terms involved, evaluation principles, working procedures, evaluation methods and technical requirements. Furthermore, we should strengthen the application of the LLEI, carry out effectiveness assessments and dynamic management in a timely manner, identify problems and deficiencies, compare differences horizontally, compare processes longitudinally, dynamically adjust and optimize construction tasks, resolve weak links in construction, jointly improve the "Liangshan" practice, and achieve green development goals.

We should also strengthen practical exploration and reinforce demonstration and guidance. We should create a number of models for the coordinated promotion of high-level protection and high-quality development, with a view to improving the supply capacity of high-quality ecological products, innovating the value realization mechanism of ecological products, exploring the transformation path of "Lucid Waters and Lush Mountains are Invaluable Assets", and to promoting ecological prosperity for local people. For regions with different resource endowments, functional positioning and development stages, we will optimize the layout of "Lucid Waters and Lush Mountains are Invaluable Assets" practice and innovation bases, deepen the construction objectives and tasks, strengthen the summary of practice models and typical cases, and provide more replicable, referable and scalable experience models.

Finally, we should strengthen incentives and constraints to stimulate innovation vitality. In light of environmental public sentiment, ecological environment monitoring network and daily supervision work, as well as central ecological environmental protection supervision and other activities, we should strengthen daily management, establish and improve supervision and exit mechanisms, and promote the quality of demonstration construction. In addition, we should strengthen support for policy funding, such as transfer payments for national key ecological functional zones and renewable energy development funds, and encourage local governments to tilt in favor of pilot areas in terms of ecological compensation, transfer payments, and Party and government performance evaluation, so as to fully stimulate innovation vitality and momentum.

Attachment: Indicators and Its Interpretation of LLEI

LLEI Evaluation Indicators Table

Target	Task	SN	Indicator	Target reference value
Building lucid waters and lush mountains	Environmental quality	1	Proportion of days with good ambient air quality	> 90%
		2	Rate of water quality of centralized drinking water sources meeting the standard	100%
		3	Proportion of surface water quality meeting or outperforming Grade III water standard	> 90%
		4	Proportion of underground water quality reaching or outperforming Grade III water	Steady increase
		5	Safe utilization rate of contaminated farmland	> 95%
		6	Safe utilization rate of contaminated land	> 95%
	Ecological regime	7	Forest and grass coverage	Mountainous areas > 60%; Hilly areas > 40%; Plain areas > 18%
		8	Species richness	Steady increase
		9	Area of eco-environmental conservation red lines	No decrease
		10	Gross ecosystem production per unit of land area	Steady increase

(continued)

H. Zhang et al., *Theory and Practice of "Lucid Waters and Lush Mountains Are Invaluable Assets" in China*, https://doi.org/10.1007/978-981-97-7340-4

(continued)

Target	Task	SN	Indicator	Target reference value
Promoting "Liangshan" transformation	People's well-being	11	Proportion of residents' per capita output value of ecological products	Steady increase
	Ecological economy	12	Output value of green and organic agricultural products as a proportion of total agricultural output value	Steady increase
		13	Output value of the ecological processing industry as a proportion of total industrial output value	Steady increase
		14	Ecotourism income as a proportion of total output value of the service industry	Steady increase
	Eco-compensation	15	Eco-compensation revenue as a proportion of total fiscal revenue	Steady increase
	Social benefits	16	International and domestic ecological cultural brands	Obtain
		17	Public satisfaction with the results of "Liangshan" construction	> 95%
Building long-term mechanisms	Institutional innovation	18	Institutional construction for "Liangshan" bases	Establishment and implementation
		19	Market-based mechanism for ecological products	Establishment and implementation
	Fund guarantee	20	Investment in ecological and environmental protection as a proportion of GDP	> 3%

Explanation of the indicators

Proportion of days with good ambient air quality

Indicator explanation: This refers to the days on which the air quality in an administrative region meets or outperforms the secondary standard as a proportion of days effectively monitored in the whole year. It follows the *Ambient Air Quality Standard* (GB 3095-2012) and the *Ambient Air Quality Index (AQI) Technical Regulations (Trial)* (HJ 633-2012).

$$\text{Proportion of days with good ambient air quality} = \frac{\begin{array}{l}\text{Number of days with ambient air quality meeting}\\ \text{or outperforming the secondary standard}\end{array}}{\text{Number of days effectively monitored in the whole year}} \times 100\%$$

Source Ecology and environment departments.

Rate of water quality of centralized drinking water sources meeting the standard

Indicator explanation: This refers to the number of centralized drinking water sources in an administrative region whose surface water quality meets or outperforms the Grade III standard of the Surface Water Environmental Quality Standard (GB 3838-2002), and underground water quality meets or outperforms the Grade III standard of the Underground Water Quality Standard (GB/T 14848-2017) as a percentage of the total number of water sources in that region.

Source Ecology and environment departments.

Proportion of surface water quality meeting or outperforming Grade III water standard

Indicator explanation: This refers to the proportion of water quality of nation-, province-, and city-controlled monitoring sections in an administrative region that meets or outperforms Grade III water standard. It follows the *Surface Water Environmental Quality Standards* (GB 3838–2002), which mandates that the surface water in an administrative region meet the standard of the water environment functional zone; the proportion of Grade I and II water quality be not decreased; and that the water quality of the passing river sections controlled by the level above the city not be reduced.

$$\begin{array}{l}\text{Proportion of surface water quality meeting or outperforming}\\ \text{Grade III standard}\end{array} = \frac{\begin{array}{l}\text{Number of sections whose water quality of nation-, province-, and}\\ \text{city-controlled monitoring sections meet or outperform}\\ \text{Grade III standard}\end{array}}{\text{Total number of monitoring sections}} \times 100\%$$

Source Ecology and environment departments and water resources departments, etc.

Proportion of underground water quality meeting or outperforming Grade III standard

Indicator explanation: This refers to the number of monitoring sites whose underground water quality meets or outperforms the Grade III standard as a proportion of the total number of monitoring sites in an administrative region. Underground water quality is divided into five categories according to the *Underground Water Quality Standard* (GB/T 14848–2017), ensuring that the proportion of water quality meeting or outperforming Grade III standard is stable or improved, and that the overall underground water quality is stable or shows a trend towards continuous improvement.

$$\begin{aligned}&\text{Proportion of underground water quality meeting or outperforming}\\&\text{Grade III standard}\\&=\frac{\begin{array}{l}\text{Number of monitoring sites whose underground water quality}\\\text{meets or outperforms Grade III standard}\end{array}}{\text{Total number of monitoring sections}}\times 100\%\end{aligned}$$

Data Source Natural resources departments, water resources departments, ecology and environment departments, etc.

Safe utilization rate of contaminated farmland

Indicator explanation: This refers to the safe utilization area of contaminated farmland as a proportion of the area of contaminated farmland in an administrative region. It follows the *Calculation Method for the Safe Utilization Rate of Contaminated Farmand (Trial).*

$$\begin{aligned}&\text{Safe utilization rate of contaminated farmland}\\&=\frac{\text{Safe utilization area of contaminated farmland}}{\text{Total area of contaminated farmland}}\times 100\%\end{aligned}$$

Data Source Agriculture and rural affairs departments, ecology and environment departments, etc.

Safe utilization rate of contaminated land

Indicator explanation: This refers to the area of polluted land for redevelopment and utilization that meets the requirements of soil environmental quality for planned land use as a proportion of the total area of polluted land for redevelopment and utilization in an administrative region.

$$\begin{aligned}&\text{Safe utilization rate of contaminated land}\\&=\frac{\begin{array}{l}\text{Area of polluted land for redevelopment and utilization}\\\text{that meets the requirements of soil environmental}\\\text{quality for planned land use}\end{array}}{\text{Total area of polluted land for redevelopment and utilization}}\times 100\%\end{aligned}$$

Source Ecology and environment departments.

Forest and grass coverage

Indicator explanation: This refers to forest and grassland area as a proportion of the total land area in an administrative region.

$$\text{Forest and grass coverage rate (\%)} = \frac{\text{Forest area (sq.km)} + \text{Grassland area (sq.km)}}{\text{Total land area (sq.km)}} \times 100\%$$

Data Source Forest and grass departments, natural resources departments, agriculture and rural affairs departments, etc.

Species richness

Indicator explanation: This refers to the number of species in an administrative region, which is an important indicator to reflect biodiversity protection.

Data Source Forest and grass departments, natural resources departments, agriculture and rural affairs departments, ecology and environment departments, etc.

Area of eco-environmental conservation red lines

Indicator explanation: This refers to an area with special important ecological functions within the scope of ecological space, which must be subject to mandatory and strict protection. It is the bottom line and lifeline of safeguarding and maintaining national ecological security. It usually includes areas with important ecological functions such as water conservation, biodiversity maintenance, soil and water conservation, wind prevention and sand fixation, and marine ecological stability, as well as ecologically sensitive and vulnerable areas such as those with soil erosion, land desertification, stony desertification, and salinization. It is required to establish an ecological protection red line system to ensure that the area of ecological protection red lines is not decreased, the nature is not changed, and that the leading ecological function is not reduced.

Data Source Natural resources departments, ecology and environment departments, etc.

Gross ecosystem production per unit of land area

Indicator explanation: This refers to the gross ecosystem production per unit of land area in an administrative region, which is an important indicator reflecting the operating status of the ecosystem in the region, and an important representation of lucid waters and lush mountains. The gross ecosystem production refers to the economic value of the products and services provided by the ecosystem for human survival and wellbeing, which mainly includes the value of the ecosystem supply service, the value of the ecosystem regulation service, the value of the ecosystem support service, and the value of the ecosystem cultural service.

$$\text{Gross ecosystem production per unit of land area} = \frac{\text{Gross ecosystem production}}{\text{Total land area}}$$

Data Source Natural resources departments, development and reform commissions, ecology and environment departments, etc.

Proportion of residents' per capita output value of ecological products

Indicator explanation: This refers to income from realization of the value of ecological products per capita of residents as a ratio of the total income per capita of residents in the statistical scope of an administrative region, which is an important indicator to measure the benefits of residents from the transformation of lucid waters and lush mountains into invaluable assets. This indicator value is assessed based on the average value of the indicator value obtained by the sampling survey of statistical departments or the sampling questionnaire survey of independent investigation institutions. The output value of ecological products refers to the value that can be directly transformed by the ecological products provided by the ecosystem, mainly including the income from agriculture, forestry, husbandry, aquatic products, carbon sink trading, and from farming and fishing tourism.

$$\text{Proportion of residents' per capita output value of ecological products} = \frac{\text{Per capita income of residents from the realization of the value of ecological products}}{\text{Residents' per capita gross income}} \times 100\%$$

Data Source Statistics departments, agriculture and rural affairs departments, forestry and grass departments, culture and tourism departments, or independent survey agencies.

Output value of green and organic agricultural products as a proportion of total agricultural output value

Indicator explanation: This refers to the contribution rate of the output value of green and organic agricultural products to the total agricultural output value in an administrative region, which is the main indicator reflecting the development status of green and organic agriculture. Green and organic agricultural products shall comply with relevant national certification provisions, with products covering planting, fishery, understory industry and husbandry, such as green and organic agricultural products planting and breeding, and Chinese medicinal materials planting. The environmental status of the producing area of green and organic agricultural products shall meet the national environmental protection standards and management requirements of the *Standards for Environmental Quality Evaluation of the Producing Areas of Edible Agricultural Products* (HJ/T 332-2006) and the *Standards for Environmental Quality Evaluation of the Producing Areas of Greenhouse Vegetables* (HJ/T 333-2006).

$$\text{Output value of green and organic agricultural products as a proportion of total agricultural output value} = \frac{\text{Output value of green and organic agricultural products of the year}}{\text{Total agricultural output value of the year}} \times 100\%$$

Data Sources Agriculture and rural affairs departments, forestry and grass departments, industry and information technology departments, ecology and environment departments, statistics departments, market supervision departments, etc.

Output value of the ecological processing industry as a proportion of total industrial output value

Indicator explanation: This refers to the contribution rate of the output value of the ecological processing industry to the total industrial output value in an administrative region, which is the main indicator reflecting the development status of the ecological processing industry. Specifically, the ecological processing industry mainly includes agricultural and sideline food processing, food manufacturing, beverage manufacturing, wood processing, furniture manufacturing, and mineral water production derived from ecological resources.

$$\text{Output value of the ecological processing industry as a proportion of total industrial output value} = \frac{\text{Output value of the ecological processing industry of the year}}{\text{Total industrial output value of the year}} \times 100\%$$

Data Source Industry and information technology departments, development and reform commissions, etc.

Ecotourism income as a proportion of total output value of the service industry

Indicator explanation: This refers to the contribution rate of ecotourism income to the total output value of the service industry in an administrative region, which is the main indicator reflecting the development status of the ecotourism industry. Ecotourism refers to a tourism mode that takes sustainable development as its concept, ecological and environmental protection as its premise, and human-nature coordination as its criterion. Relying on a good natural ecological environment and a unique humanistic ecosystem, ecotourism carries out ecological experiences, ecological education, and ecological cognition in an environmentally friendly manner, satisfying tourists physically and mentally.

$$\text{Ecotourism income as a proportion of total output value of the service industry} = \frac{\text{Ecotourism income of the year}}{\text{Total output value of the service industry of the year}} \times 100\%$$

Data Source Culture and tourism departments, industry and information technology departments, development and reform commissions, etc.

Eco-compensation revenue as a proportion of total fiscal revenue

Indicator explanation: This refers to the contribution rate of fiscal revenue from eco-compensation to total financial revenue in an administrative region, which is the main indicator reflecting the transformation of ecological environmental protection effectiveness. The fiscal revenue from eco-compensation includes central transfer payments, provincial transfer payments, and horizontal compensation between regions.

$$\text{Eco-compensation revenue as a proportion of total fiscal revenue}$$
$$= \frac{\text{Fiscal revenue from eco-compensation of the year}}{\text{Total fiscal revenue of the year}} \times 100\%$$

Data Source Finance departments, ecology and environment departments, etc.

International and domestic ecological cultural brands

Indicator explanation: This refers to international and domestic brand identities obtained in an administrative region that prominently feature environmental friendliness and ecological advantages and that have ecological cultural connotations, ecological cultural influence and additional appeal. They mainly involve world cultural heritage, world natural heritage, the United Nations "Champion of the Earth Award", national intangible cultural heritage, national civilized cities, national ecological civilization construction demonstration zones, national forest cities, and national ecological tourism demonstration zones.

Data Sources Industry and information technology departments, ecology and environment departments, agriculture and rural affairs departments, culture and tourism departments, etc.

Public satisfaction with the results of "Liangshan" construction

Indicator explanation: This refers to the degree of public satisfaction with the effectiveness of the construction of "Liangshan" in an administrative region. This indicator is obtained by a combination of random questionnaires issued on site by the "Liangshan" base construction evaluation working group and sampling surveys commissioned to an independent authoritative public opinion survey institution. The average value of the indicator values obtained by the field survey and those obtained by the independent survey institution is the final result. The number of respondents of the field survey shall be no less than one thousandth of the population of an administrative region. The respondents shall include people of different ages, different educational backgrounds and different occupations to be considered fully representative.

Data Sources Questionnaire survey and sampling survey by an independent institution.

Institutional construction for "Liangshan" bases

Indicator explanation: This refers to promotion by an administrative region of the construction of "Liangshan" bases through institutional construction, including establishing organizational leadership and operational mechanisms to coordinate base construction; establishing a system for the breakdown and implementation of the construction objectives of "Liangshan" bases, clarifying work objectives, work tasks, time nodes, strengthening the sense of responsibility, and ensuring the steady progress of work.

Data Source Local governments and relevant departments.

Market-based mechanism for ecological products

Indicator explanation: This refers to the policy mechanism established in an administrative region around the market-based trading and operation of ecological products and their value. Referring to the Overall Plan for Structural Reform in Ecological Civilization, the market-based mechanism for the transformation of ecological products primarily includes the trading system on energy use rights and carbon emission rights, the trading system on emission rights, the trading system on water rights, and the green finance system.

Data Sources Development and reform commissions, finance departments, natural resources departments, ecology and environment departments, water resources departments, agriculture and rural affairs departments, forestry and grass, etc.

Investment in ecological and environmental protection as a proportion of GDP

Indicator explanation: This refers to the annual investment in ecological environmental protection, governance or maintenance as a proportion of gross regional domestic production (GDP) in an administrative region, which is the most intuitive indicator to reflect the degree of importance attached to the ecological environment and the status of ecological environmental governance in the construction area of "Liangshan" bases.

$$\begin{aligned}&\text{Ecological and environmental protection investment as a proportion of GDP}\\&=\frac{\begin{array}{l}\text{Investment in ecological and environmental protection,}\\\text{governance or maintenance of the year}\end{array}}{\text{GDP of the year}}\times 100\%\end{aligned}$$

Data Sources Development and reform commissions, industry and information technology departments, natural resources departments, ecology and environment departments, and statistics departments, among other departments.

References

1. Almond R E A, Grooten M, Petersen T. Living Planet Report 2020 - Bending the curve of biodiversity loss[R]. WWF, Gland, Switzerland, 2020.
2. Boulding K E. The economics of the coming spaceship earth[R]. Baltimore: Johns Hopkins University Press, 1966.
3. Chen Demin. The core connotation of circular economy is the recycling of resources [J]. China Population, Resources and Environment, 2004, 14 (2):12–15.
4. Chen Zhongxin, Zhang Xinshi. The value of ecosystem benefits in China [J]. Chinese Science Bulletin, 2000, 45:17–22.
5. China Research Center for Sustainable Development, Peking University (CRCSD). The Path of Sustainable Development [M]. Beijing: Peking University Press, 1994.
6. Feng Zhiming, Yang Yanzhao, Yan Huimin, et al. Research on resource and environmental carrying capacity over the past century: from theory to practice [J]. Resources Science, 2017, 39 (3):379–395.
7. Fu Bojie. The development trend and priority areas of ecosystem research in China [J]. Geographical Research, 2010, 29 (3):383–396.
8. Fu Yun, Ma Yonghuan, Liu Yijun, et al. Research on the development model of low-carbon economy [J]. China Population, Resources and Environment, 2008, 18 (3):14–19.
9. Grossman G M, Krueger A B. Environmental impact of North American free trade agreement, the US Mexico free trade agreement [M]. P Garber, Cambrige, MA: MItPress, 1993.
10. IPCC. Climate Change 2021: The Physical Science Basis [M]. London: Cambrige University Press, 2021.
11. Li Gang. COP15: Building a global consensus to reverse biodiversity loss [N/OL]. Yangcheng Evening News, 2021-10-18. http://news.youth.cn/hotnews_41880/202110/t20211018_13266360.htm.
12. Ouyang Zhiyun. Problems and countermeasures faced by China's ecosystem [J]. China National Conditions and Strength, 2017(3):6–10.
13. Pan Jiahua, Zhuang Guiyang, Zheng Yan, et al. Concept identification and core elements analysis of low-carbon economy [J]. International Economic Review, 2010, 4:88–102.
14. Pan Jiahua. Providing value-added dividends of ecological products [N/OL]. Economic Information Daily, 2017-10-23, http://www.jjckb.cn/2017-10/23/c_136698923.htm.
15. Pan Yue. How to supply more high-quality ecological products [N]. People's Daily, 2018-01-31 (020).
16. Plumptre A J, Baisero D, Belote R T, et al. Where Might We Find Ecologically Intact Communities?[J]. Frontiers in Forests and Global Change, 2021(4):1–13.
17. Qi Hongqian, Wang Zhitao. The development logic and trend characteristics of ecological economics [J]. China Population, Resources and Environment, 2016, 26 (7):101–109.
18. Qu Geping. Developing circular economy is the general trend of the 21st century [J]. China Environmental Protection Industry, 2001 (S1):4–5.

H. Zhang et al., *Theory and Practice of "Lucid Waters and Lush Mountains Are Invaluable Assets" in China*, https://doi.org/10.1007/978-981-97-7340-4

19. Rees W E. Ecological footprints and appropriated carrying capacity: What urban economics leaves out [J]. Environment and Urbanization, 1992, 4(2): 121–130.
20. Ripple W J, Wolf C, Newsome T M, et al. World Scientists Warning to Humanity: A Second Notice [J]. Bioscience, 2017. https://doi.org/10.1093/biosci/bix125.
21. Sachs J D, Lafortune G, Fuller G. The SDGs and the UN summit of the future. Sustainable development report 2024 [R]. Paris: SDSN, Dublin: Dublin university press, 2024.
22. Strassburg B B N, Iribarrem A, Beyer H L. et al. Global priority areas for ecosystem restoration [J]. Nature, 2020(586):724–729.
23. UK Government. Our Energy Future: Creating a Low Carbon Economy [R]. UK Government Energy White Paper, 2003.
24. United Nations Environment Programme (2021). Making Peace with Nature: A scientific blueprint to tackle the climate, biodiversity and pollution emergencies. Nairobi. https://www.unep.org/resources/making-peace-nature.
25. Xi Jinping. Promoting China's ecological civilization construction to a new level [J/OL]. Qiushi, 2019 (3). http://www.qstheory.cn/dukan/qs/2019-01/31/c_1124054331.htm.
26. Yan Lidong. The concept and characteristics of eco-economic system[J]. Chinese Journal of Eco-Agriculture, 1996(2):27.
27. Yan Xin, Research on residual evolution trend and driving mechanism of low-carbon economy in China [D]. China University of Geosciences (Beijing), 2020.
28. You Fei, Wang Chuansheng. Discussion on basic theory, research method and discipline development trend of ecological economics [J]. China Soft Science, 2003, 3:131–138.
29. Zhang Huiyuan. Study on major practice models and evaluate index of "Lucid waters and lush mountains are invaluable assets" in China [J]. Environment and Sustainable Development, 2020(6): 104–107.
30. Zhang Qiang, Liu Yujie, Zhang Huiyuan, et al. Path analysis of ecological civilization governance capacity building [J]. Environment and Sustainable Development, 2015, 40 (4): 10–14.

SPRINGER NATURE

GPSR Compliance

The European Union's (EU) General Product Safety Regulation (GPSR) is a set of rules that requires consumer products to be safe and our obligations to ensure this.

If you have any concerns about our products, you can contact us on ProductSafety@springernature.com

In case Publisher is established outside the EU, the EU authorized representative is:

Springer Nature Customer Service Center GmbH
Europaplatz 3
69115 Heidelberg, Germany

The manufacturer's authorised representative in the EU is Springer Nature Customer Service Centre GmbH, Europaplatz 3, 69115 Heidelberg, Germany. If you have any concerns regarding our products, please contact ProductSafety@springernature.com

Printed and bound by CPI Group (UK) Ltd, Croydon, CR0 4YY
30/09/2026
02225265-0007